1/10/74

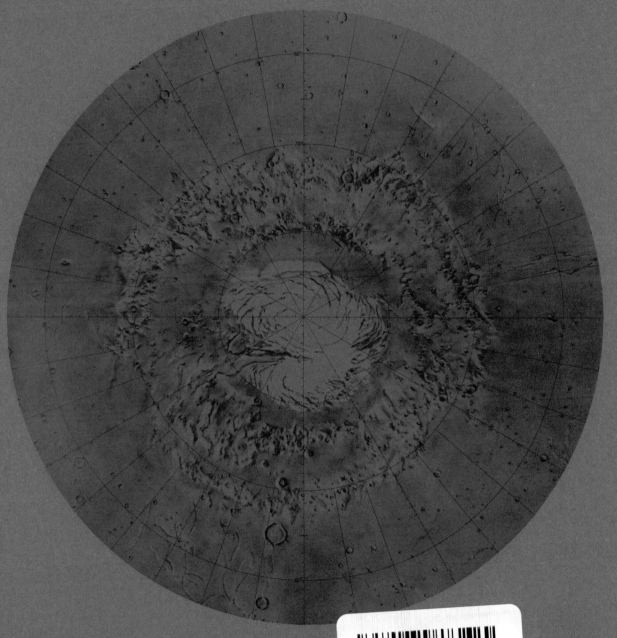

NORTH POLAR RE

D1223366

Dutchess Community College

Library

Poughkeepsie, N. Y.

Mars and the Mind of Man

Mars and

the Mind of Man

Ray Bradbury

Arthur C. Clarke

Bruce Murray

Carl Sagan

Walter Sullivan

1817

Harper & Row, Publishers
New York, Evanston, San Francisco, London

523.43082
M363

Portions of this book appeared in *Horizon*.

MARS AND THE MIND OF MAN. Copyright © 1973 by Harper & Row, Publishers, Inc. All rights reserved. Printed in the United States of America. No part of this book may be used or reproduced in any manner whatsoever without written permission except in the case of brief quotations embodied in critical articles and reviews. For information address Harper & Row, Publishers, Inc., 10 East 53rd Street, New York, N.Y. 10022. Published simultaneously in Canada by Fitzhenry & Whiteside Limited, Toronto.

FIRST EDITION

Designed by Patricia Dunbar

Library of Congress Cataloging in Publication Data

Main entry under title:
Mars and the mind of man.

  Panel discussion held in Nov. 1971 at the California Institute of Technology, with R. Bradbury, A. C. Clarke, B. Murray, C. Sagan, and W. Sullivan; with afterthoughts written after Oct. 1972.
  1. Space flight to Mars. 2. Project Mariner. 3. Mars (Planet)—Exploration. 4. Astronautics and civilization. I. Bradbury, Ray, 1920–
TL799.M3M37   508.99′23   72–9746
ISBN 0–06–010443–0

Contents

Preface

As preparations for Mariner 9's arrival at Mars in November 1971 reached their climax, an unusual assemblage of personalities was brought together in Pasadena by their interest in the planet. Thus was born the idea of a public discussion about Mars and the Mind of Man. As a Caltech faculty member, I was the obvious one to organize and host the panel.

And so it has been also with this book, except that I have had the aid of the Harper & Row staff. Frances Lindley contributed significantly to the organization and structure of the text. Patricia Dunbar transformed the manuscript and individual photographs into a visual feast far surpassing the authors' expectations.

An acknowledgment is in order for Caltech itself. The radical revision of scientific and popular thought about Mars which has taken place recently is almost entirely a product of the flights of Mariner spacecraft to Mars. That challenging enterprise has been made possible through the extraordinary dedication and skill of the engineers of the Jet Propulsion Laboratory. More broadly, we are all indebted to the wisdom of Caltech over past decades for encouraging such a noble technological goal for JPL, one abounding in intellectual and cultural promise, yet devoid of military secrecy or other inhibitions to free communication.

In the final analysis, wise leadership, engineering skill, and scientific imagination were focused toward major achievement only because planetary exploration has appealed to the imagination of the American people for over a decade. May that enlightened and enthusiastic spirit continue as our hallmark.

BRUCE C. MURRAY

Pasadena, California
January 1973

Foreword: On Going a Journey
by Ray Bradbury

When I was very young, twelve years old, a carnival used to come to my small town in northern Illinois every Labor Day weekend. In that carnival was a traveling illusionist and onetime-but-now-defrocked Presbyterian minister (so he said!) named Mr. Electrico.

Mr. Electrico, for some reason still unknown to me—perhaps my own ebullient loud spirits called to him, perhaps he lacked a son anywhere in the world—took me on as once-a-year friend. I looked forward to his brief return each autumn, for then we would walk along the Lake Michigan shore behind the carnival and talk great philosophies (mine) and small philosophies (his). (And if you think I jest, consider how philosophies dwindle instead of swell with your own age.)

Anyway, it was during one of these strolls that Mr. Electrico (his real name has long since vanished with the autumns) revealed to me that we had met previously, long years before I was born. That is always great news to a twelve-year-old boy! To think one has already lived once or twice in this strange world, and to have an older man tell you of a collision of souls far beyond remembrance? Delicious.

Where had we met? On the battlefield at the Argonne in France during World War I. I had died in his arms in that Argonne conflict. He had seen my soul flicker out in the eyes of another man, that same soul to be rekindled and named Ray Bradbury on a summer day in late August 1920.

Well, of course, I thought Mr. Electrico was a pip and loved him so well that I named a character for him in my novel *Something Wicked This Way Comes.*

What has all this to do with the book you have in your hands? How does it lead us to Mars and our thinking of Mars?

The answer comes as simply as going out to gaze at that

Red Planet shining in the sky on a proper night. I have always looked on myself as some sort of Martian. My affinity for the planet is immense and prolonged and most affectionate-fine. And if I could have died once on the battlefield at the Argonne in the arms of a defrocked theologian, well then, I might easily have once lived on Mars, and amiably choose to believe so.

I have also chosen to start this book in this fashion, for soon enough it will get serious and you will find yourself up to the gills with facts. And facts quite often, I fear to confess, like lawyers, put me to sleep at noon. Not theories, however. Theories are invigorating and tonic. Give me an ounce of fact and I will produce you a ton of theory by tea this afternoon. That is, after all, my business. And, come to think of it, the business of the men—most of them with a good sense of humor, thank God—who have allowed me to shiv in here at the front and along the way through the book.

Anyway, we have established my claim to other lives, probably on other planets, which gives me the hubris to attempt, without pontificating, this Foreword.

I make the claim of Martian, also, because it is a biological/theological point I will return to again and again and again as the book progresses. We are all Aristotle's children, which is to say children of the Universe. Not just Earth, or Mars, or this System, but the whole grand fireworks. And if we are interested in Mars at all, it is only because we wonder over our past and worry terribly about our possible future.

And even if we never were Martians in our deep, dark root years of prehistory, the day is fast coming when we shall name ourselves so.

I foresaw this (not smugly, I hope) when, some twenty-three years ago, I wrote a strange tale entitled "Dark They Were, and Golden-Eyed."

In that Martian story, I told of a man and his family who helped colonize Mars, who eat of its foods and live in its strange seasons, and stay on when everyone else goes back to Earth, until the day finally comes when they find that the odd weathers and peculiar temperatures of the Red Planet have melted their flesh into new shapes, tinted their skin, and put flecks of gold into their now most fantastic eyes, and they move up into the hills to live in old ruins and become—Martians.

Which is the history I predict for us on that far world. The ruins may not be there. But if necessary we will *build* the ruins, and live in them and name ourselves as my transplanted Earthmen did. And will not be of Earth any more but will truly be Martians, just as in the not far future we will be Moon Creatures and then, God and Time willing, benevolent circumnavigators of an as yet unselected target-sun.

We are, then—at this moment, because we dream it so—Martians. We wish to be that thing and so will be it.

And this book is one of the Openers of the way to that old dream, now refurbished and put forth in metals and fires to establish as a profound twentieth-century truth what seemed a fantasy.

In all this I feel like a twelve-year-old boy lost among politicians, or worse, the mob which pelts me with stones and rocks and cries "escape" as label to my dreaming and to space travel. It is not escape I am interested in. We have been suffering a crisis of the spirit for fifty years, one hundred years, two hundred years, perhaps longer. Man does not need escape so much as he needs release into a new spirit, a transcendent knowledge of himself that only Space can give him.

If the Moon was one large step for mankind, Mars is the next largest.

I speak then, here, of Going a Journey.

I borrow the title from Hazlitt, who spoke of the joys of traveling alone through countryside, with blue sky above, green turf below, and one's thoughts to amble with.

So man, in our times, goes a journey, and the destination is very far and at present has no name, and we indeed travel alone, for mankind is the lone thing; nothing like him exists in our part of the universe, and our thoughts are long and sometimes filled with that joy which brims itself to terror.

What of that journey, the rocket and its meaning, man and his endless ticketing of himself to Far Rockaway and Lands End and Copernicus Crater? Will we never get him away from the Viking longboat, off the trolley, out of the jet, free of the rocket or the damn time machine he so dearly wishes to invent, test, explode, and go far-traveling with?

Never.

Will any of it improve him?

About as much as ten laps around a meadowfield and a cold shower help a boy of fifteen. It doesn't change him; it but makes him feel more alive.

How can you possibly compare space travel with a sweaty boy and an icy shower?

Because I want mankind to be very *much* alive. But *improve* him? No. Hitler and Stalin wanted to improve him out of existence.

I would take him—warts, bumps, hogwash, mush, and all, every athlete's foot of him, armpit lumps, corns, bad dreams—and put him on the Moon, Mars, then drop him in the Coal Sack Nebula shouting with joy, shrieking with fear, and alive, alive, O!

I don't think you can improve a thing that is already improved, already lost; always behind but always winning; filled with midnight, burning with sun; sly and untrusty, open and lacking guile.

I sing paradoxical man.

I accept not only his flesh but the bones within his flesh and the sin marrowing those bones.

Approve of him? It is hard to approve of this lumpy child. But sons are always lovable, murderers though they be, saints though they be—and we hate saints sometimes, do we not, as much as we hate murderers?

I sing the entire man, then, going into Space.

We must therefore know ourselves better, which means adding intelligence to intelligence, brick to brick. That great void, that gaped wound in the side of God which bleeds itself to ignorance, must be stopped up, filled in with the stuffs that we give it, item by item, data following data, as man fleshes out himself and binds up the dying away of his Progenitor. God births us even as we must, in the living, birth Him.

But I sound like a stepped-on lion. Let the great caterwaul subside. I am a born teacher, which means push my button and a drowsy lecture ensues. Forgive. I cannot stop up that young boy who still stomps around inside me yelling wonders.

I will let the book, its pictures, and my amiable confreres take over. You will have to come back and step on my tail again later and put up with a few more ignorant yowls and semi-theological half-aesthetic screeches. You will be worn out with forgiveness by the time I have wrung myself dry.

Here are four other good chaps. Behind and beyond them, Mars. Mars speaks with a great voice. If you listen closely, here on Earth, my friends are trying to be heard.

Gentlemen, please to begin.

1 Hypotheses

NOVEMBER 12, 1971

Mariner 9 is close to its historic rendezvous with Mars. Tomorrow the spacecraft is scheduled to fire its braking rocket and be captured by the gravity field of Mars. Once in orbit, its cameras and other scientific instruments are intended to map systematically the entire surface of the planet. It is anticipated that twenty or thirty times as many pictures and other scientific data as were sent by the three previous Mariners will be transmitted by Mariner 9.

Man should at last be able to discover the true identity of his planetary neighbor. Is Mars a brother of the Earth, as was generally believed before the first Mariners in 1965 and 1969? Or is it nothing more than a cousin of the Moon, as the results of the early Mariners seemed to indicate? Or can it be a member of an as yet unknown family? What are the two small natural satellites of Mars like?

In Pasadena, at Caltech's Jet Propulsion Laboratory, engineers are listening carefully to the radio signals returned from the spacecraft. Is their robot completely ready for its complicated task? A whole task force of scientists—astronomers, geologists, physicists, chemists, and meteorologists—are busy readying their final plans for the incredibly complex set of individual tasks they wish Mariner 9 to execute.

On the Caltech campus, a group of five articulate and thoughtful men, brought together by Mariner 9's arrival at Mars, and by their deep interest in the planet, are sitting down for an encounter uniquely their own. Two are famous science fiction writers, Ray Bradbury and Arthur C. Clarke, who have come on a kind of sentimental journey to find out what the imaginary planet they wrote of is really like with her make-up off. Two are leading authorities on Mars, Bruce Murray and Carl Sagan, who are taking a break from their work with that busy scientific task force. Walter Sullivan, Science Editor for *The*

New York Times, is here to cover Mariner 9's arrival for his newspaper.

What follows is the edited transcript of their encounter.

Sullivan

Clarke   Bradbury

Sagan   Murray

Walter Sullivan: "on the eve of turning another page in the history of man's understanding of the planetary system in which we reside"

SULLIVAN: I'd like to outline the operating rules for the panel discussion. After I've taken about ten minutes to introduce our subject, each speaker will have about ten minutes to express some of his thoughts on it. Then we'll have about a twenty-minute discussion with the group here and then we'll have another twenty minutes, if our timetable holds up, for questions from the audience.

The subject of Mars and the mind of man is most appropriate today, on the eve of turning another page in the history of man's understanding of the planetary system in which we reside. At least I hope we're going to turn another page late tomorrow afternoon.* I think that some of our friends here on the platform have as strong and as authoritative views about what we're going to find on that page as anybody in the world and it will be very exciting to hear some of their thoughts on this. The central subject is this problem of Mars in the mind of man. In itself this is a fascinating subject, for which reason it is appropriate that we have two science fiction writers who are very scientific as well as fictional.

My mind harks back to before we knew very much about the planetary environments and it seemed rather logical to assume that the planets were all inhabited. There was a time when this belief was not confined to the kooky fringe. People as illustrious as Immanuel Kant and others of this period believed that all the planets were inhabited and that the temperament of the inhabitants was determined by their distance from the sun.

* November 13, 1971, the scheduled time for Mariner 9 to fire its braking rocket.

5

In other words, the inhabitants of Mercury were mercurial, hot-tempered, fierce, ferocious, very uncivilized types. Inhabitants of Jupiter, which was so far from the sun, had a very cool and serene temperament. Well, of course, as we got to know more and more about the planets, this possibility became less and less probable and the chances narrowed down to Venus and Mars. We couldn't see the surface of Venus, so there was talk of an oceanic planet full of sea monsters and all kinds of other things.

But we could see that there was geography or Martianography on the surface of Mars, and as the better telescopes and man's imagination came into play, the hopes of there being a supercivilization on Mars of course grew by leaps and bounds. This viewpoint achieved its climax very recently in the life span of astronomy. It was within the memories of some of us up here on the platform, at least one or two. In 1924, there was the closest opposition* of this century, just a little bit closer than the one this summer. Now, this was an extraordinary period. The astronomers had begun to realize that the environment of Mars was very, very inhospitable to life, at least as we knew it. But there was such a popular desire, one might say, that there should be a civilization on Mars even superior to our own that public pressure persuaded both the Chief of Naval Operations and the Director of the U.S. Army Signal Corps to send dispatches to their stations to maintain, insofar as possible, radio silence during the opposition in case the Martians try to communicate to Earth with their more advanced radio technology. One astronomer—I'd hate to say he was at

---

* Mars is in opposition to Earth when both are aligned on the same side of the sun. At this time, the distance between the two planets is minimal. Such oppositions occur every twenty-five months. Due to the eccentricity of the orbits of both planets, the minimum distance between the two planets at opposition varies. The opposition of 1924 was an especially favorable time for viewing Mars.

Cornell,* I think he was at Williams—was told that there was a mine shaft in Chile pointing naturally to the zenith, being vertical, and he calculated that every night during the opposition Mars would pass right over the top of this mine shaft. So it was his proposal to install a spinning disc at the bottom of this mine shaft and fill it with mercury, the idea being that the mercury would then spin into a paraboloid mirror about seventy feet in diameter and this would bring the Martians within two miles and we could see what they look like. It was an interesting idea but didn't get very far.† Nevertheless, the Army was persuaded to bring in its chief cryptographer, a man named William Friedman. In those days, in 1924, nobody had ever heard of William Friedman. He was to make history, more than a decade later, by breaking the Japanese code. He was, indeed, our best cryptographer! At the recent international meeting on extraterrestrial intelligence that was held in the Soviet Union, I am told, the Russians had one of their best cryptographers present to discuss the problem of trying to decipher a message from some other intelligent civilization at great distance, a message not designed to be undecipherable but quite the opposite—a message designed to be decipherable by some intelligence that had no other basis on which to operate except logic, whatever that is.

In fact, the concept of the Martians is still so strong, at least in the American mind, that on Halloween of this year a radio station in Buffalo broadcast a program that they had actually recorded earlier. It was a modern version of the famous one of 1938, when Orson Welles adapted the H. G. Wells program of *The War of the Worlds* to have Martians landing in New Jersey. This year, however, they landed in Grand Island, a suburb of Buffalo, and this radio station was very enterprising.

* Carl Sagan is Professor of Astronomy at Cornell University.
† Atmospheric turbulence—not aperture—limits the resolution of telescopic observations of Mars from Earth.

It had its mobile unit out there describing the scene. It had several reporters scattered around describing the panic-stricken population fleeing in all directions. This program had been promoted weeks in advance with press accounts announcing that it would be a fake, but still the police switchboard lighted up. There were so many people who believed in Martians!

So, let us undertake to discuss today what our friends here on the platform think about Mars and the public mind, and *in* the mind of man.

Carl Sagan: "There is no question that the straightness of the lines is due to intelligence. The only question concerns which side of the telescope the intelligence is on."

SULLIVAN: We'll start out with a gentleman, on my left, whom I have known for a great many years. I first knew him when he looked, if possible, younger than he does today and he had already built a reputation for being both precocious and provocative in the field of planetary science. His pungent remarks tended to raise many eyebrows. He also became interested in a subject in which I'm interested too—the possibility of there being intelligent life in very distant worlds and the analysis of this possibility. He took the book by Iosif Shklovski of the Soviet Union on this subject, Shklovski being one of the great theoretical astrophysicists alive today, and together they produced an expanded version of this same book. Carl is now the editor of *Icarus*, I believe. It has a marvelous subtitle—something like "International Journal of the Solar System." I don't know how many subscribers they have on Mars, but they have a great many on Earth! And as one of his side jobs (being facetious) he is also Director of the Laboratory for Planetary Studies at Cornell: Carl Sagan.*

SAGAN: As I understand my job in these few minutes, it is to trace, even if briefly, some of the flavor of the early discussions on Mars within the decades around the turn of the century, a time that tended to mold opinions about Mars from

* Sagan was on leave from Cornell at this time to participate as a member of the Television Team of Mariner 9. He has been a principal exponent of the search for life on Mars and is usually associated with the more optimistic view concerning the environment and possible biological history of Mars.

then to the present. I myself first became aware that Mars was a place of some interest by reading stories by Edgar Rice Burroughs, who also is known for his invention of Tarzan. Burroughs invented a gentleman adventurer from Virginia named John Carter, who was able to transport himself to the planet Mars by standing in an open field and sort of spreading his arms out and wishing. At least that's as close as I could get to the method. And at an early age, eight or nine, I tried very hard to put the Carter method to the experimental test. But no matter how hard I tried, it failed—perhaps not entirely to my surprise, but I thought there was always a chance. And so now, by proxy, we're going to Mars at this moment but not in nearly so interesting a way.

The Mars that Burroughs imagined he called Barsoom. In fact, he had a lovely phrase, "the hurtling moons of Barsoom," which of course are the two moons of Mars, Phobos and Deimos. The first (indistinct) close-up photograph of Deimos was made yesterday by Mariner 9. Barsoom was a dying planet. It had drying canals and races of ancient peoples. Now, where did those ideas come from?

Classically, the first impetus for a dying Mars comes from the nebular hypothesis of Kant and Laplace, a view of the origin of the solar system not too different from what is fashionable today. A vast gas and dust cloud of interstellar dimensions contracts and spins up, as it contracts, to conserve angular momentum. As escape velocity is reached in the equatorial plane, little blobs of matter get spun off progressively from the outer regions of the solar system inwards, each of which, by some process not investigated in detail by Kant and Laplace, condensed into a planet. That meant that the outermost planets were older and the innermost planets were younger. Mars was accordingly older than the Earth, and Venus younger. If you believed that the time interval of formation was significant compared to the age of the solar system, Mars might be significantly older, Venus significantly younger than the Earth. You might imagine Mars as a dying Earth and Venus as the Earth hundreds of millions of years ago. Today we know that the time interval for the formation of the planets was very short compared to the lifetime of the solar system. The planets cannot be of very different ages.

The observational basis for this idea of Mars as a dying world was provided first by an Italian astronomer named Giovanni Schiaparelli but was publicized consummately by an American brahmin from Boston, diplomat to Korea turned astronomer, named Percival Lowell. He was the brother of the president of Harvard and of Amy Lowell, the poet. Lowell advocated observing Mars from a place where the atmosphere was reasonably steady (or, as the astronomer says, where the "seeing" is good). Then by eyeball astronomy you look through the telescope and you draw pictures of what you see. Lowell was unfortunately one of the worst draftsmen who ever sat down at the telescope and the kind of Mars that he drew was composed of little polygonal blocks connected by a multitude of straight lines. These were the straight lines that had first been reported by Schiaparelli in 1877, when there was an opposition of Mars very much like the present one. He called these lines "canali," which in Italian means channels or grooves. But it got translated as canals, and you can see the whole hypothesis was right there in the translation. Somebody saw canals on Mars. Well, what does that mean? Well, canal—everybody knows what a canal is. How do you get a canal? Somebody builds it. Well, then there are builders of canals on Mars. Fundamentally, Lowell's argument was that no natural process could produce such a network of long straight lines; hence, they were artificial; hence, there were artisans.

I would like to give some of the flavor of the debate on the canals. Here are a few sentences from Lowell. Even then, astronomers knew that Mars had much less water than does the Earth. Lowell says, "The fundamental fact of the matter is the dearth of water. If we keep this in mind we shall see that many of the objections that spontaneously arise answer themselves. The supposed herculean task of constructing such canals disappears at once; for, if the canals could be dug for irrigation purposes, it is evident that what we see, and call by ellipsis the canal, is not really the canal at all, but the strip of fertilized land bordering it, the thread of water in the midst of it, the canal itself, being far too small to be perceptible. In the case of an irrigation canal seen at a distance, it is always the strip of verdure, not the canal, that is visible, as we see in looking from afar upon irrigated country on the Earth." This is in response to one of the more basic objections to the ideas of canals—namely, that they would be too small to see. The syntax, I think, is as interesting as the substance.

The basic idea was that there were canals constructed by a race of vast intelligence on Mars to channel the waters from the melting polar caps to the thirsty inhabitants of the equatorial cities of Mars. Since there wasn't a lot of water around, you had to be careful about conserving it. Now, there are two questions. One: are there such features on Mars? Two: if they are on Mars, need they be due to the interpretation that Lowell put on them? So let us turn to some other planetary astronomers.

The first statement is by E. E. Barnard in 1894. "I have been watching and drawing the surface of Mars. It is wonderfully full of detail. There is certainly no question about there being mountains and large, greatly elevated plateaus. To save my soul, I can't believe in the canals as Schiaparelli [or Lowell] draws them. I see details where he has drawn none. I see details where some of his canals are but they are not straight lines *at all*. When best seen, these canals are very irregular and broken up—that is, some of the regions of his canals; I verily believe, for all the verifications—that the canals as depicted by Schiaparelli are a fallacy and that they will be proved so before many oppositions are past."

And then a second skeptical remark, by E. M. Antoniadi. "At the first glance through the 32¾-inch telescope 1909, September 20, I thought I was dreaming and scanning Mars from his outer satellite. The planet revealed a prodigious and bewildering amount of sharp or diffused natural irregular detail all held steadily; and it was at once obvious that the geometrical network of single and double canals discovered by Schiaparelli was a gross illusion. Such detail could not be drawn; hence, only its coarser markings were recorded in the notebook." These two latter descriptions conform very nicely to what we now know of the appearance of Mars. The canals of Mars are probably due to the eye's penchant for order. It is much simpler to draw disconnected fine detail as a few lines, joining them up, than to put down all the irregular mottlings observed in an instant of good seeing. There is no question that the straightness of the lines is due to intelligence. The only question concerns which side of the telescope the intelligence is on. Lowell appreciated this point perfectly well: "The straightness of the lines is unhesitatingly attributed to the draughtsman." Now, this is a very telling point, he says, "For it is a case of the double-edged sword; accusation of design, if it proves not to be due to the draughtsman, devolves *ipso facto* upon the canals." And then, in words we can all take to heart, he concludes, "Let us not cheat ourselves with words. Conservatism sounds finely, and covers any amount of ignorance and fear."

Well, this is the highest level of the pro-canal polemic. There are other levels. Here are a few lines from a book called *World Making,* by Samuel Phelps Leland, Ph.D., Ll.D., Emeritus Professor of Astronomy in Charles City College and author of

Peculiar People, Etc., published in Chicago in 1898 by the Women's Temperance Publishing Association. He says, "When next the planet and the Earth come in opposition great discoveries will then be made. The planet will be high in the heavens. The telescope of the Chicago University with its 40″ glass will probably then be completed. The telescope will almost double the space penetrating power of the 36″ refractor at Mt. Hamilton" (a slight mathematical error: $40^2/36^2$ does not equal 2). Then comes the terrific part: "With such a power it will be possible to see cities on Mars, to detect navies in its harbors and the smoke of great manufacturing cities and towns. And it may be possible to flash electrical signals across the space which could be readily seen by the inhabitants of Mars with telescopes of considerable power and the answer easily seen by us." And then he concludes, "Is Mars inhabited? There can be little doubt of it. His [that is, Mars'] conditions are all favorable for life and life of a high order. It is not improbable that there are beings there with a civilization as high if not higher than our own." Then, in a lovely use of words, he says, "Is it possible to know this of a certainty? Certainly."

I want now to get to what I think is the high point of the intellectual discussion on Mars in this period. You can see that there were discussions at not always the highest intellectual level, but surely the idea of life on Mars was very exciting. There was one man who looked at it from a vantage point not of the professional astronomer, not of the professional publicist, and not of the science fiction writer, but from a very different view. This man was Alfred Russel Wallace. He was the co-discoverer, with Charles Darwin, of evolution by natural selection. He was off in Sumatra for decades as an old-time naturalist. He sent a paper to Darwin to be transmitted to the Linnean Society which had the whole theory of Darwin's work right in it. He was a very clever fellow. Well, Wallace was asked to review a book of Lowell's. His book review, written in a white heat, was itself of book length. I have it here, *Is Mars Habitable?,* published in 1906. He attacked Lowell on physical and not biological grounds. He discovered an error in Lowell's estimate of the albedo of the Earth and correctly deduced a mean temperature of Mars of about 230°K,* well below the freezing point of water. Lowell thought that Mars had a temperature comparable to the south of England, apparently the temperature standard of the time. Wallace believed that the annual temperature variation was extreme, that the polar ice caps were at least in part made of condensed carbon dioxide, that the surface material was porous, that craters were to be found in abundance on the surface of the planet, that large amounts of water vapor were not to be found because of gravitational escape, that the canals, if they existed, were related to geological faults, and that Phobos and Deimos were residua of the formation of the planet. He was just on the verge of deducing subsurface permafrost on the planet. The book was published in Wallace's eighty-third year; he died shortly thereafter.

Reading Wallace's book, I am just astounded at the excellence of his logical powers and the currency of many of his conclusions. He has occasional lapses, as in his conclusion that Mars is more like the Moon than the Earth.† Also, the fact that he thought there was no water at all led him to conclude in the last sentence of the book, "Mars not only is not inhabited by intelligent beings as Mr. Lowell postulates but is absolutely" (and the last word is capitalized) "UNINHABITABLE." By that he means by large organisms.

Well, this is a flavor of what was happening when ideas about Mars were for the first time being widely discussed.

* Water freezes at 273°K on the absolute or Kelvin scale.
† A view associated with Murray and other scientists concerned with interpreting the television pictures returned by Mariners 4, 6, and 7.

After Wallace, the debate passed from scientific works through Sunday supplements to science fiction writers' minds, and then spread to a vast public to establish the popular views of what Mars was like. I hope in the discussion later I'll have a chance to say what I think Mars is really like.

SULLIVAN: I might say, I was trying to park my car and suddenly everything sort of vanished all around. I was in a parking lot across the way. Arthur was landing in a helicopter!

CLARKE: I guess *I* was kicking up some dust as well. Incidentally, I do *think* this dust cloud on Mars is extremely suspicious!

I want to go along with Ray Bradbury here. It was Edgar Rice Burroughs who turned me on, and I think he is a much *underrated writer*. The man who can create the best-known character in the whole of fiction should not be taken too lightly! Of course, there's not much left of his Mars, and his science was always rather dubious. I can still remember even as a boy feeling there was something a little peculiar about cliffs of solid gold, studded with gems. I think it might be an interesting exercise for a geology student to see how that phenomenon could be brought about.

Another writer I'd like to pay tribute to, partly because he lived such a tragically short time, was Stanley G. Weinbaum, whose *Martian Odyssey* came out around 1935. And then, of course, the other great influence on me was our Boston brahmin. Whatever we can say about Lowell's observational abilities, we can't deny his propagandistic power, and I think he deserves credit at least for keeping the idea of planetary astronomy alive and active during a period when perhaps it might have been neglected. He certainly did a lot of harm in some ways, but I think perhaps in the long run the benefits may be greater.

Anyway, I was very moved the other day when I visited the Lowell Observatory for the first time and actually looked through his 26-inch telescope. He's buried right beside it; his tomb is in the shape of the observatory itself. I was distressed to find that his papers had been rather neglected and scattered around. As a result, I have initiated a series of events which may now result in his papers being classified and, hopefully, edited. Whatever nonsense he wrote, I hope that one day we will name something on Mars after him, and I'm sure that he won't be forgotten in this area.

And then, of course, you mentioned H. G. Wells. He certainly did a lot for Mars, and is still doing so, as you heard the other day. I don't know if movie director George Pal is here in the audience, but he has also done a lot for Mars, not to mention Los Angeles, in *The War of the Worlds.* He gleefully destroyed City Hall and a few other places around here.

We are now in a very interesting historic moment with regard to Mars. I'm not going to make any definite predictions, because it would be very foolish to go out on a limb, but whatever happens, whatever discoveries are made in the next few days or weeks or months, the frontier of our knowledge is moving inevitably outwards.

It has already embraced the Moon. We still have a great deal to learn about the Moon and there will be many surprises even there, I'm sure. But the frontier is moving on and our viewpoint is changing with it. We're discovering, and this is a big surprise, that the Moon, and I believe Mars, and parts of Mercury, and especially space itself, are essentially *benign* environments—to our technology, not necessarily to organic life. Certainly benign as compared with the Antarctic or the oceanic abyss, where we have already been. This is an idea which the public still hasn't got yet, but it's a fact.

I think the biological frontier may very well move past Mars out to Jupiter, which I think is where the action is. Carl, you've gone on record as saying that Jupiter may be a more hospitable home for life than any other place, *including* Earth itself. It would be very exciting if this turns out to be true.

I will end by making one prediction. Whether or not there is life on Mars now, there *will* be by the end of this century.

Rebuttals and Responses

SULLIVAN: Now we can proceed to a question period. First, I'd like to follow up Bruce Murray's challenge. Carl, do you have any comment on the suggestion that perhaps we don't really need to sterilize the Viking spacecraft so elaborately and expensively? This has been a subject of concern for many, many years. There was an organization called CETEX. Among its activities was an international project to ensure sterilization of all spacecraft that were going to land on other celestial bodies where there might be life. But there has been a certain, I think, lack of unanimity between the Americans and the Russians on this subject. At least, there has been a strong suspicion that the Russians did not believe that heat sterilization was necessary. I think they use a sterilizing gas. So Carl, what do you think?

SAGAN: One of the main points Bruce has stressed is that our desires or wishes may influence our decisions and conclusions. I think that's very true and very human. Perhaps a similar case is the whole UFO business, where the wish is father to the observation, at least in some cases. But because a possibility is interesting, it doesn't follow that it's false. We can be emotionally predisposed as pessimists as well as optimists. Actuarial procedures provide a guide to situations of this sort. How careful you have to be in a given situation and how much premium you have to pay is not only a question of how likely the event in question is but also how important the event is. Suppose, for example, we're concerned about carrying terrestrial microorganisms to Mars, depositing them there, and having them survive and multiply so that the next generation of space vehicles finds the next generation of microbes. How do we then distinguish Earth life from Mars life? If that's what we're worried about, then it's not enough to say that the survival of terrestrial organisms on Mars is unlikely. We also have to worry about how much damage is done if, despite the unlikelihood, terrestrial contamination of Mars occurs. And it is the product of these two points, the likelihood and the importance, which is what determines, in my mind, the necessity of sterilizing Mars-bound space vehicles.

There is no question that the Martian environment is hostile for terrestrial forms of life in the most parochial sense. Yet there are wide variations. For example, a commonly discussed impediment to terrestrial life on Mars is the solar ultraviolet flux, which is terrifically intense. In fact, a resistant terrestrial microorganism sitting on the surface of Mars at an ordinary time gets UV-fried in about one second. It just shrivels up and dies. But a microorganism on Mars at this particular moment doesn't have that problem. By accident, there happens to be a large dust storm that's obscuring the surface. The ultraviolet absorption by the dusty atmosphere is much larger than for visible light. Now is a terrific time for Martian organisms, if there are any, or terrestrial contaminants (which aren't there yet) to start moving around on Mars. There are also possibilities of near-surface liquid water. The chances of contaminating Mars are small, but they're not negligible. Arthur's last remark was, "There'll certainly be life on Mars by the end of this century." I'd like to add, "Especially if we don't sterilize the space vehicles."

On whether the Soviets are interested in heat sterilization, from talking to Soviet scientists, my understanding of the situation in Soviet sterilization of space vehicles is the following: The Russians blanch at the idea of having electronic circuitry which will go through temperature soaks much above the normal boiling point of water. My guess is, and I may be wrong, that the two Soviet space vehicles, each of which I believe will contain an entry probe, have been thoroughly surface-sterilized by gaseous sterilization and radiation and heat. The insides may also have been presterilized by some of these methods. Alternatively, the inside of those space vehicles is loaded with bugs, but the inside of those space vehicles is also loaded with thermite. The spacecraft enters the Martian atmosphere, survives to the surface, does whatever it has to do on the surface, and then, on command from the Earth, the thermite is ignited and the solution of the three-dimensional equation of heat conduction then proceeds. Even those micro-organisms in the thermally most inaccessible parts of the spacecraft are fried many times over. The spacecraft does not blow open and every bug is killed. Whether such a scheme will work or not is another question. But on the question of whether the Soviets take the question of sterilization seriously, I think the answer is certainly yes.

Can I say a word about the question of life on Mars? This is a different question from what we've just been talking about. Is it possible that there is life on Mars, Martians? Now, just as there have clearly been excesses in the direction of prematurely concluding that there is life on Mars, and I have quoted some of those excesses, I think there have also been excesses in the other direction, in prematurely concluding there isn't life on Mars. We have a certain intolerance for ambiguity, saying, "Don't confuse me with the facts, just give me an answer." Well, I think that's where we are on the question of life on Mars. There is, as far as I can tell, no more reason to conclude that Mars is lifeless than there is to conclude that it is inhabited. There is water, there is carbon dioxide, there is sunlight—these are the prerequisites even for parochial forms of green plant photosynthesis.

The likelihood of life arising in the early history of Mars is, I think, well within the bounds of possibility. We have no observations which give us strong evidence one way or another about the early history of Mars or this question of the origin of life there, and it certainly is not out of the question that Mars does have organisms today. I don't think Arthur or Ray should be apologizing quite so prematurely, although I'd be terribly surprised if the *Martian Chronicles* scenario comes through.

Let me conclude my remarks on indigenous life on Mars by asking the following question: At what point in our exploration of Mars could we detect ourselves? That is, suppose we took the most optimistic assumption, namely, that Mars has a civilization on it of precisely the terrestrial extent and development at the present time. Would we have detected it? An interesting question, which calibrates how far along we are in the biological exploration of Mars. Now, there's a trivial way in which we would have detected it. Just as we are beaming all sorts of radio noise into space which can be detected—as housewives' daytime TV serials and other low forms of intelligence—in the same way, were there a civilization precisely of our extent on Mars, we would be detecting it in the radio spectrum. But, of course, there was no radio emission a hundred years ago, when the Earth was populated by intelligent beings, and there may be none a hundred years from now, when I hope the Earth will still be so populated, because of closed circuit and cable TV. So I don't consider the absence of TV from Mars an important criterion.

What about photographic detection? If we were just going in to ask whether there is life on Earth, then by looking at the resolution of the pictures, and how many of them you have to look at in order to detect life on Earth, the answer comes out quite clearly: with the number of photographs taken of Mars up to the present time, with the resolution we've used, were it the other way around, we would not have detected life on Earth. And an interesting point is that the first mission which has a hope of detecting life on Mars of contemporary terrestrial extent is Mariner 9. At the end of Mariner 9, all the observation curves start bumping into the detection curves for discovering our terrestrial civilization. Now, I don't think there is an advanced civilization on Mars, for statistical reasons. But we cannot exclude it. The remarkable thing is that Mariner 9 is the first mission which gives us the capability to test such a hypothesis. And surely, simpler forms of life on Mars cannot be excluded by the photographic detection method that has been developed up to the present time. So I don't think there is cause for optimism about life on Mars, but I don't think there is cause for breast-beating either. I think the proper attitude is to keep an open mind and see what the observations uncover. Mariner 9 is not designed to look for the more likely varieties of life on Mars and I shall be very surprised if it gives us any convincing evidence one way or another.

SULLIVAN: In case there is anyone in the audience who hasn't heard the story of the first astronaut to return after Arthur Clarke's prediction— When he finally got back, he stepped from the spacecraft on the carrier deck and the people rushed up and said, "Is there life on Mars?" And he said, "Well, you know, it's pretty dead most of the week, but it really swings on Saturday night."

Now I'd like to go back to Bruce Murray—he's been scribbling notes there—and ask him whether, if we buy his argument that the probability of life is really that low, is Viking worthwhile?

MURRAY: I think there is a lot of gum stuck on the side of that baseball you just pitched to me. It's a screwball!

SULLIVAN: No screwballs here! But to rephrase my question, at least is the life-detection component of the Viking lander worthwhile?

MURRAY: I think that as long as looking for life on Mars is the objective, it is necessary to go directly to the surface. There is surely no way to find out about life without taking direct measurements. Any system capable of landing and remotely carrying out something as complicated as a biological experiment will be very expensive, and very complicated. Furthermore, the most likely outcome of such an effort is failure to prove the existence of life there. Even with Carl's optimistic estimates it's a long shot—perhaps ten to one. At worst, it's a million to one. Nobody is going to give you even money or anything like that. So, even if everything works right, the chance for success of the life-detection experiments is low.

On the other hand, the *desire* of the American people (who are paying the costs) to search for life on Mars has been high, and Viking is a logical translation of that desire into a space mission. I guess I would chalk up that desire and enthusiasm in the face of what I consider unpromising odds to Lowell and Edgar Rice Burroughs, and to Ray Bradbury and Arthur Clarke. Hence, Viking is a response to a genuine public interest. We've come so far on this life-on-Mars kick, we can't turn back now, even if the recent scientific evidence has not been encouraging for life.

SULLIVAN: A footnote question I want to ask now is: "How many landings do we have to make on Mars with negative results before we can say there is no life on Mars?"

MURRAY: I've pressed my life-on-Mars colleagues pretty hard on this question. They are quite disagreed among themselves. It is quite clear that one landing with negative results will not be sufficient to cause their scientific opinion to swing around entirely. There will be those who will say that you didn't look in the right place, or at the right time, or in the right way. From my personal view, the question of looking for life on Mars with automated landers may be a modern version of the pursuit of the Golden Fleece. Even if there is some peculiar microbial life there, we'll never know for sure with such primitive robots as Viking, expensive as they are. I think the only practical way to look for life is to bring a sample back (with automated spacecraft) and analyze it in laboratories on Earth. I think the lunar experience with the Apollo samples is a beautiful illustration of that. The knowledge we gained of the Moon from the returned sample is enormously greater than anything done remotely on the surface.

SULLIVAN: It's the old problem that it's very easy to say yes when you have some definite evidence, but to say no confidently is very difficult. Let me ask Ray Bradbury whether he feels the concept of Mars in the mind of man—this great emotional desire to find life there—is a good thing or a bad thing.

BRADBURY: I think it is especially good. It's fascinating to see here on the platform today how many start as romantics, and in fact hate to give it up. I think it's part of the nature of man to start with romance and build to a reality. There's hardly a scientist or an astronaut I've met who wasn't beholden to some romantic before him who led him to doing something in life.

I think it's so important to be excited by life. In order to get the facts we have to be excited to go out and get them, and there's only one way to do that—through romance. We need this thing which makes us sit bolt upright when we are nine or ten and say, "I want to go out and devour the world, I want to do these things." The only way you start like that is with this kind of thing we are talking about today. We may reject it later, we may give it up, but we move on to other romances then. We find, we push the edge of science forward, and I think we romance on beyond that into the universe ever beyond. We're talking now about Alpha Centauri. We're talking of light-years. We have sitting here on the stage a person who has made the film with the greatest metaphor for the coming billion years. That film is going to romance generations to come and will excite the people to do the work so that we can live forever. That's what it's all about. So we start with the small romances that turn out to be of no use. We put these tools aside to get another romantic tool. We want to love life, to be excited by the challenge, to live at the top of our enthusiasm. This process enables us to gather more information. Darwin was the kind of romantic who could stand in the middle of a meadow like a statue for eight hours on end and let the bees buzz in and out of his ear. A fantastic statue standing there in the middle of nature, and all the foxes wandering by and wondering what the hell he was doing there, and they sort of looked at each other and examined the wisdom in each other's eyes. But this is a romantic man —when you think of any scientist in history, he was a romancer of reality. . . . You see, when you ask a short question, you get a long answer.

SULLIVAN: A beautiful answer. My mind was going back to mythology, the beginning of all of that, when they looked to the sky and they wove their myths of gods and things in with the stars and planets they saw. But Arthur, he's thrown the ball to you.

CLARKE: Walter, your remark about the value of Mars to us reminds me of an answer Jim Van Allen* gave when someone said, "What is the use of the Van Allen Belts?" He said, "Well, I make a pretty good living out of them. . . ." To get back to the question of life on Mars—or anywhere else, for that matter —we are learning that the chemicals of life are far more widespread than we had ever dared to imagine. Who'd have dreamed that such complex organic molecules exist in meteorites, as my friend Cyril Ponnamperuma† has been discovering? And there's pretty good reason now for thinking that, given half a chance—or even one chance in ten—or even in a million— life not only evolves, but evolves very rapidly. Of course, we're also pretty sure that there will never be anything exactly like life on this earth elsewhere, because there are so many random

* Dr. James Van Allen of Iowa State University is credited with one of the first major discoveries of the Space Age in 1958, the intense belts of trapped radiation that surround the Earth.
† A well-known Ceylonese biochemist at the University of Maryland, who has experimented with the production of early organic compounds and, with others, looked for them in meteorites.

choices; all the different chance throws of the genetic dice will never produce the same thing twice, except in an infinite universe.

I quite agree with Carl Sagan that we have probably switched too far to the other extreme. When I said that Mars (I'll argue about Mercury and the Moon some other time, so let's stick to Mars) is a benign environment, I was thinking of technology, but I wouldn't rule out that adjective even for biological evolution. If life ever had a chance of getting started on Mars, it could still flourish there. We forget that Mars is a very small planet, with no oceans; it also has a long year, so any reasonably mobile life form could probably always remain in a pretty optimum condition—it only has to migrate a mile or so a day. It could always enjoy an endless summer, to coin a phrase. And I'd also like to try to shoot down the idea that if there are any life forms on Mars, they must be primitive. I would say the exact reverse—they're going to be damned sophisticated. I think it would be a good idea to watch out—they might be hungry for oxygen and carbon and hydrogen—and heat!

SULLIVAN: We have time now for a few questions from the audience.

QUESTION: The evidence seems to show no life on Mars. But maybe the life there is much more advanced than here, and they've all left their bodies. You know, pure spirits. What if you find that?

SAGAN: Bruce Murray will be glad to know that the person whose views correspond most closely to his advocates spiritualism!

Well, we don't have good statistics on how many forms of life there are. On Earth, there's only one kind. All the organisms on earth are fundamentally the same kind. Organisms like beetles and begonias and people all look different, but they're identical in terms of biochemistry. So I would be content to find a slight variation, never mind one that doesn't have a body or whatever—some slight variation in the chemistry or the nucleic acids or enzymatic catalysis from what we have here. That would be pretty exciting for me. But if someone doing an astronomical observation of Mars bumped into a spirit there, I suppose he would submit his findings to the *Astrophysical Journal* in the usual way.

SULLIVAN: Carl Sagan always argues that we shouldn't be so provincial, so parochial, in our concepts of life. I wonder if he's familiar with J. B. S. Haldane's proposal that there might in fact be some kind of silicate biological activity that could occur deep within the earth? This conjures up all kinds of wild ideas—you remember Conan Doyle's story of the deep-well diggers in Scotland who dug deeper and deeper until they hit something soft and squashy . . .

How about some more questions from the audience?

QUESTION: When will we or the Russians really start looking for life on Mars?

SAGAN: As far as I understand it, there are no "life detectors" on Mars 2 or Mars 3, which are now on their way a little bit behind Mariner 9, in any sense other than the sense that Mariner 9 instrumentation is life-detection, which it isn't. But Mariner 9 and Mars 2 and 3 data might tend to establish boundary conditions for life. As far as I can see, the Soviets will not be landing "life detectors" on the Martian surface before we will, around 1976.

I cannot resist speaking up on Walter's comment about silicon-based life. I think that's a red herring that runs through the semiscientific literature on the Red Planet. Carbon molecules just seem better for biology than silicon molecules. But there is one kind of experiment that isn't based upon assumptions about Martian biochemistry, and that's an imaging system. You land cameras on Mars, and if a silicon-based giraffe walks by, you've got it! Or if you have a seismic system, and the elephants are big enough . . .

QUESTION: What is the latest on the hypothesis which I. S. Shklovski published in your joint book, *Intelligent Life in the Universe,* that Phobos and Deimos are artificial satellites of Mars?

SAGAN: The very latest is that Deimos was photographed by Mariner 9 yesterday. But the story is like this: In the *Astronomical Journal* in about 1944 there is a paper by B. P. Sharpless, who worked for the U.S. Naval Observatory. He published a reduction of all the data on the motions of the moons of Mars from 1877, when they were first observed. His reduction of the data indicated a secular acceleration of the inner moon of the same sort that artificial satellites have when they decay into the Earth's atmosphere. End of Sharpless's paper. Then several other people tried to explain this motion and failed. Then Shklovski approached the problem by proposing a large range of alternatives, say 1 through 37—none of which worked. Now the reason the usual satellite drag explanation doesn't work is that the Martian atmosphere is so rarefied that there isn't enough drag to produce the supposedly observed secular acceleration. Shklovski said, well, maybe the moon isn't solid, maybe it doesn't have all that mass, so the little tiny bit of atmosphere is able to drag it down anyway. He calculated what the mass had to be, then calculated what the density had to be. It turned out the thing had to be hollow. So, well, that's very interesting. Here's this thing orbiting Mars—it's ten miles across and it's hollow; what must it be? You can't avoid concluding it's an artificial satellite launched by a species of very considerable technological attainments. There does not seem to be any evidence of such a species on Mars today; therefore, there must have once been an advanced civilization on Mars. That's the end of Shklovski's argument. The argument isn't bad. The trouble is with the observations. Because now

G. A. Wilkins in Britain finds that there is no good evidence for the secular acceleration and Shklovski has withdrawn his hypothesis. But maybe Mariner 9 will get a good close-up photograph of Phobos, and really check the idea out.*

SULLIVAN: We have time for one more question.

QUESTION: Mr. Bradbury, do you have any more poems?

BRADBURY: I thought you would never ask! In the last few years, I have found myself returning again and again to the problem of science and theology. This problem has thrust itself into the center of a series of poems I have written. I have for some time now thought that the conflict between religion and science was a false one, based, more often than not, on semantics. For when all is said and done, we each share the mystery. We live with the miraculous and try to interpret it with our data correctors or our faith healers. In the end, survival is the name of the game.

Once upon a time we created religions which promised us futures when we knew there were no possible ones. Death stared us in the face, forever and ever.

Now, suddenly, the Space Age gives us a chance to exist for a billion or two billion years, to go out and *build* a heaven instead of promising one to ourselves, with archangelic hosts, saints waiting at Gates, and God pontifical on his Throne.

This second poem of mine is titled "Old Ahab's Friend and Friend to Noah Speaks His Piece." It is written from the viewpoint of the whale speaking to future men, telling them they must build a whale and live inside it and go out into space in it and travel through time to survive forever. Here is the conclusion of that poem:

I am the Ark of Life. You be the same!
Build you a fiery whale all white.
Give it my name.

* It did. See the Epilogue section and the Mariner 9 photos of Phobos.

Ship with Leviathan for forty years
Until an isle in Space looms up to match your dreams,
And land you there triumphant with your flesh
Which works in yeasts, makes wild ferment,
Survives and feeds
On metal schemes.
Step forth and husband soil as yet untilled,
Blood it with your wives, sow it with seeds,
Crop-harvest it with sons and maiden daughters,
And all that was begat once long ago in Earth's strange waters
Do recall.
The White Whale was the ancient Ark.
You be the New.
Forty days, forty years, forty-hundred years,
Give it no mind;
You see. The Universe is blind.
You touch. The Abyss does not feel.
You hear. The Void is deaf.
Your wife is pomegranate. The stars are lifeless and bereft.
You smell the Wind of Being.
On windless worlds the nostrils of old Time are stuffed
With dust and worse than dust.
Settle it with your lust, shape it with your seeing,
Rain it with your sperming seed,
Water it with your passion,
Show it your need.
Soon or late,
Your mad example it may imitate.

And gone and flown and landed there in White Whale craft,
Remember Moby here, this dream, this time which does suspire,
This kindling of your tiny apehood's fire.
I kept you well. I languish and I die.
My bones will timber out fresh dreams,
My words will leap like fish in new trout streams
Gone up the hill of Universe to spawn.
Swim o'er the stars now, spawning man,

And couple rock, and break forth flocks of children on the plains
On nameless planets which will now have names;
Those names are ours to give or take.
We out of Nothing make a destiny,
With one name over all
Which is this Whale's, all White.
I you begat.
Speak then of Moby Dick,
Tremendous Moby, friend to Noah.
Go. Go now.
Ten trillion miles away,
Ten light-years off,
See! from your whale-shaped craft;
That glorious planet!

Call it Ararat.

2 Afterthoughts

More than a year has passed since Mariner 9 approached Mars. The five men brought together by that event, who shared their thoughts with each other and with their audience at Caltech, have gone their separate ways.

It was a year in which Mariner 9 returned 7,500 pictures and vast amounts of other kinds of scientific data about the planet before "running out of gas" (literally)* in October 1972. The Russians were not so fortunate. The lander portion of their large spacecraft Mars 3 reached the surface only to fail seconds later, and only a small amount of new information was obtained from the orbiter portion of Mars 3. The companion spacecraft Mars 2 was less successful.

How do the same five men feel now—about Mars, about Earth? How do their thoughts come out when written down in solitude rather than spoken in public exchange, when the audience is unseen and diverse rather than visible and enthusiastic?

* Mariner spacecraft must be maintained in fixed orientation with respect to the Sun and a reference star. This is accomplished by frequent release of very small pulses of nitrogen gas through attitude-control jets. When the nitrogen gas supply finally became exhausted on Mariner 9 it began to tumble uncontrollably, lost solar power, and was switched off. It thus orbits Mars silently in a state of suspended animation, perhaps to be retrieved by a future race of astronauts in a nostalgic mood.

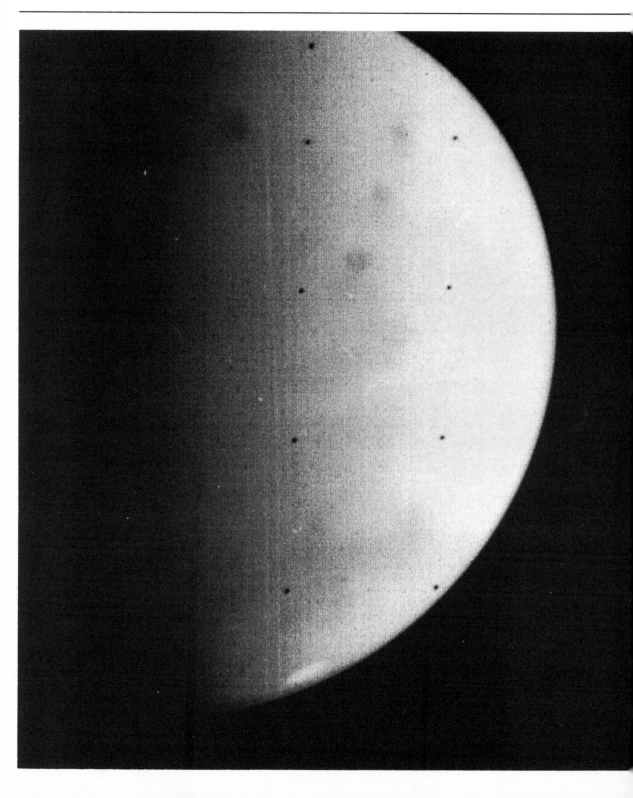

Bruce Murray

As I write these words, almost exactly a year after that high point in our lives when Mariner 9 first went into orbit about Mars, I discover somewhat to my surprise that I still seem to be playing the role of the heavy. Mars turned out to be different from what any of us thought, and demonstrated that I personally have, to some extent, been a victim of my own prejudices about the planet. Yet, I still find myself on the less optimistic side about the possibility of life on Mars, and even cautious about the promise of space itself. There can be no question that the Mariner 9 episode has been quite a trip, a peak in the history of American science, and of exploration generally. And I am deeply moved by Bradbury's poetic vision of the future in space, Arthur Clarke's stirring drama *2001,* and Carl's eloquent description of what ought to be there and why we ought to pursue it. Yet, I am not able to shake off fully the persistent realities of our present terrestrial predicament. Can the promise of space survive the despair of the cities? Will America fulfill its destiny as the leader of the Imaginative and the Good in our twentieth-century civilization? Can the obsolescence of our institutions, governmental and social, lead to constructive evolution rapidly enough so that we can capitalize on the fantastic foundations of exploration that have been laid so recently with Mariners and Apollos? I don't know the answers.

Space to me is a colorful thread intimately woven into the enormous tapestry of human existence and experience. We cannot appreciate its significance, except as part of the overall picture being traced out day by day throughout the billions of human lives on this planet. Those of us who practice our trade most directly in space exploration may perceive its potential in especially clear terms. Yet, we too are part of that tapestry and can never separate ourselves from it.

So, from this not too detached point of view, let us look

47

MARS SHROUDED BY DUST
The planet as it appeared a day and a half before Mariner 9 was put into orbit on
November 13, 1971. Dust in the atmosphere obscured all surface detail except four dark
spots near the equator and the bright south polar cap at the bottom of the picture.

backward at what Mariner 9 has learned about Mars, at how that fits into this long love affair between Mars and the mind of man, and pass on to what the future seems to hold for the exploration of Mars in particular and space in general. Finally, let's inquire a bit into how the future in space is both a mirror and an index of our future on Earth.

To me, perhaps the most surprising single aspect of the Mariner 9 mission was the discovery of huge volcanic features in the equatorial regions of the hemisphere of the planet that had not been observed by Mariners 4, 6, or 7. As these features were first observed through the dust storm, and only the gigantic calderas at their top were visible, I simply couldn't believe that they were volcanic—in fact, not only volcanic, but larger than any comparable volcanoes on the Earth. When they emerged fully from the dust, we discovered that Nix Olympica was over 300 miles in diameter and that the caldera at the crest of the volcanic mountain was larger than the *entire* island of Hawaii that stands above the Pacific Ocean. It became obvious, even to me, that Mars had developed in one place on its surface an even more Earthlike feature than the Earth.

This discovery had a double importance. First, it indicated that Mars is in a period of transition, that the rather lunar-like crust exposed throughout much of the planet, as had been observed by the first three Mariners, was being destroyed and replaced in one area by this younger volcanic terrain. I believe this process is the result of deep internal "boiling" within the planet which started up in relatively recent times. Thus, rather than having once been like the Earth and then losing its atmosphere and ending up as a dried-up apple, so to speak, Mars to me is a planet that was rather like the Moon to begin with but is now on its way to becoming Earthlike. The other important consequence, and one appropriate to the topic of Mars and the mind of man, was that when the photographic

A DUSTY LIMB
Several layers of high-altitude haze are shown detached from the main dusty atmospheric mass along the limb of the planet. In the lower right are wavelike patterns in the top of the dust, caused by the presence of high-altitude mountains on the surface of Mars.

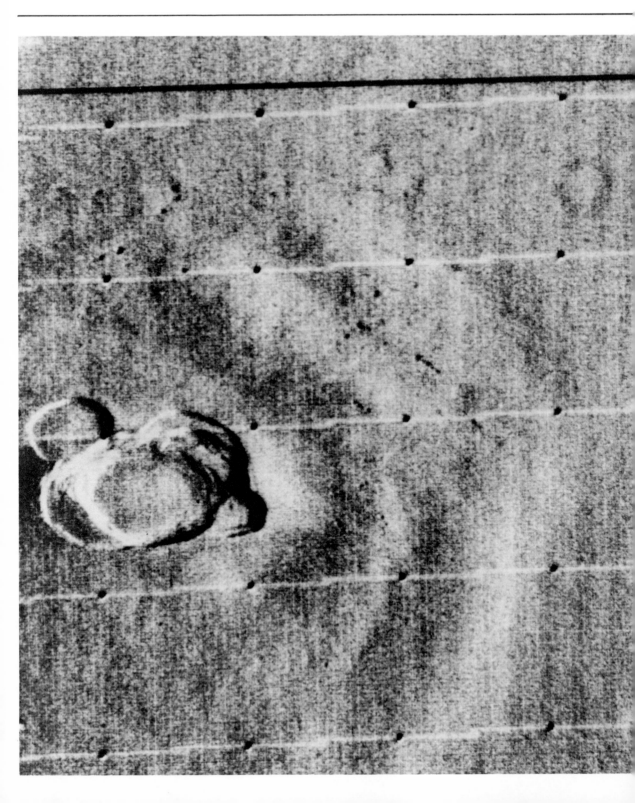

evidence of these huge volcanoes first came in, I simply couldn't accept its significance. I too was a victim of the very process I described during the panel a year ago, of being so captured by the prejudices that had grown up in my own mind about the planet as to have great difficulty in accepting and understanding the significance of new data when it arrived. Thus, my very own words at that time about a scientist's difficulty in being objective about Mars were brought home to me with stunning clarity in this particular experience.

This view of Mars as a planet in transition is supported by many other features—the canyons, the channels, the polar terrains—all of which seem to me to indicate a variety of relatively recent activities showing strong interaction by the atmosphere with the surface and creating a broad sweep of features similar to the Earth's but often on a much grander scale. I've gone so far as to speculate, perhaps in a partly contentious mood, that the very atmosphere itself may be a relatively late arrival on Mars—late in a geological sense, as within the last 1 or 2 billion years.

Thus, at least in over-all terms, I do feel that the evidence of Mariner 9 is strongly suggestive that Mars was indeed like the Moon throughout a significant fraction of its history, but being a larger planet began finally to heat up internally as a result of radioactivity, as has the Earth. This internal heating has caused the planet to begin to "boil," leading to deep convection and large-scale vulcanism at certain places, as well as the release of volatiles from its interior to form the present atmosphere and probably also a significant additional amount of frozen water-ice and solid CO_2.

As a result, Mars has proved to be an even more interesting planet to explore than I had thought a year ago, one which may well still record extraordinary episodes that took place many billions of years ago on the Earth, whose record is forever removed here by subsequent erosion and crustal de-

A FIRST LOOK AT THE NORTHERN DARK SPOT
Barely visible through the dust is a huge volcanic caldera made up of several coalesced craters. The complex is about 40 miles across and is visible here because it stands at the top of a huge volcanic mountain and protrudes through the top of the dust storm.

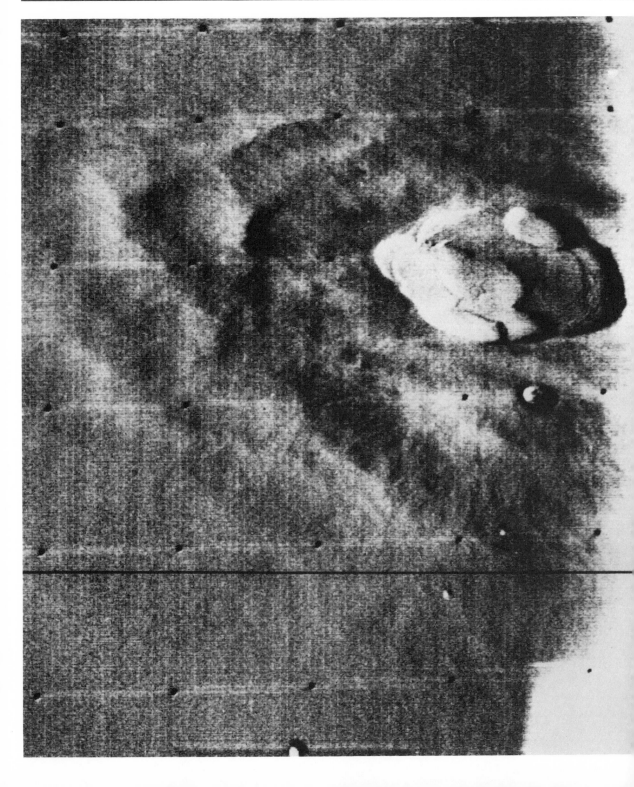

formation. The Moon has never passed through this phase. Mars may, therefore, be truly unique as an example of planetary evolution.

On the other hand, this hypothesis of an evolving Mars obviously reduces even further the likelihood that there once was an Earthlike Mars in which oceans persisted, including an Earthlike atmosphere and conditions suitable for the indigenous development of life. Even the mysterious channels seem to me to represent a brief episode in the history of the planet and not to involve massive erosion by flowing water.

Thus, Mariner 9 has perhaps filled in some of the geological history of Mars, but in such a way as to make it even more difficult to understand how life could have formed on the planet. Yet, the issue of life on Mars cannot really be joined effectively on the basis of observations from orbit: just witness the reaction of Sagan and myself to the same set of data. As each of us draws opposite conclusions about the significance of the Mariner data regarding life on Mars, it is difficult to escape the impression that we are both interpreting it in the context of differing *a priori* views of the planet.

Since Sagan and I respect each other greatly as scientists and find much stimulation in each other's thoughts, why should we find it so difficult to read the record similarly? One can look first to our scientific backgrounds. He aimed at the planets and research from his undergraduate days. My first love was —and is—the Earth, and my initial postgraduate activities were of an applied nature. I didn't return to a university for a research career until I was twenty-nine years old. Carl has emphasized synthesis and conjecture about how things are, might be, or could be beyond the Earth. If he is lucky, his great passion, the search for extraterrestrial life, especially intelligent life, will blossom during his lifetime. I, on the other hand, have been mainly concerned about distinguishing fact from fiction in a subject moldy with misconceptions and inherited prejudices.

AN EARLY VIEW OF NIX OLYMPICA
The variable bright feature called Nix Olympica by early astronomers corresponds to the upper left-hand dark spot seen in the photograph opposite page 47. It is another huge volcanic mountain.

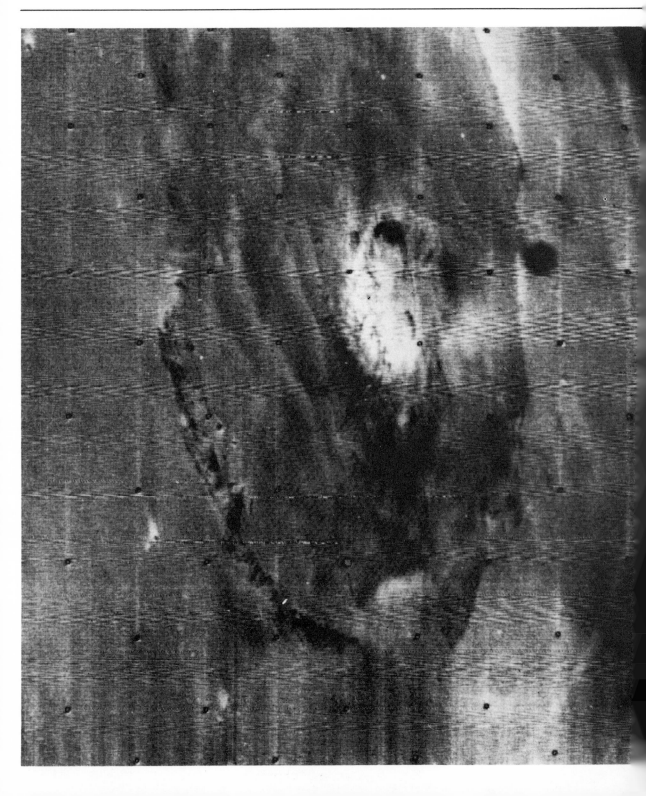

My passion is to understand how things *really* are on Earth as well as in space.

To return to the question of the biological nature of Mars, that can only be addressed by direct analysis of surface materials. Such direct analysis almost began this last time when the Russian spacecraft, Mars 3, made a successful entry and landing on the surface but survived for only about 20 seconds before mysteriously terminating transmissions. The Russians attribute this failure to very high winds associated with the raging dust storm that was taking place at that time. Had it survived, Mars 3 would have returned, I believe, excellent facsimile pictures of the surface as well as the results of some simple chemical analyses of the soil and atmosphere. It is doubtful that any experiment was aboard that could really test for life directly, but there may have been some kind of primitive environmental test of limited biological significance.

In late 1973, I expect that we will see at least two more Russian landers, this time without any American spacecraft for company, and probably at least one of these will be successful. I'll be disappointed if we don't see close-up pictures of the surface of Mars then and learn the results of some preliminary environmental measurements. In the following opposition of Mars—that is, during the early and middle parts of 1976—the American Viking landers are expected to set down with the capability for sophisticated organic analysis and for certain kinds of biological tests. There are great hopes placed on this mission by American scientists, but I personally continue to doubt that even as complex and expensive a robot as Viking is capable of carrying out so difficult a task as the unambiguous detection of life on a foreign planet by remote means. Nevertheless, the scientific harvest should be rich. It should permit much better understanding of what Mars is made up of and therefore something of the chemical history of

AN OBLIQUE VIEW OF NIX OLYMPICA

The huge shieldlike form of Nix Olympica is more apparent in this oblique view. The summit is located in the bright patch in the center of the dark field.

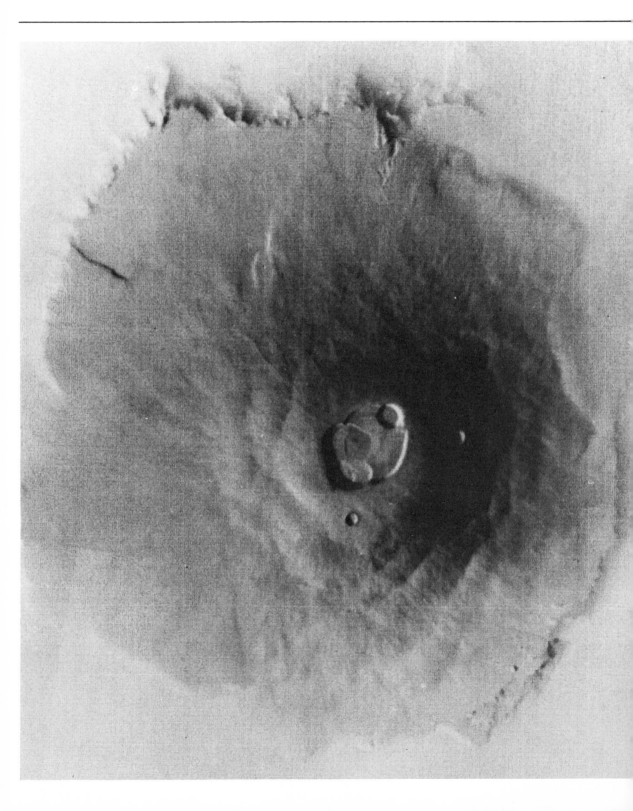

the planet, just as the Mariner pictures have led to our under-standing something of its geological surface history.

The Russians, too, can be expected to deploy a space-craft system, perhaps comparable to Viking, at that point in time. It is even possible that we may see the next step in the evolution of their program for Mars. I believe they are working toward acquiring on Mars the kind of mobility achieved on the Moon with the automatic Lunokhod. In fact, there have been some articles in the Soviet press discussing some of the difficulties of designing a Marsokhod—that is, an automatic vehicle that can traverse considerable distances on Mars, acquiring data and radioing it back to Earth. It is conceivable that our 1976 Viking lander might be accompanied by a Soviet Marsokhod.

But what next in Man's efforts to explore his fascinating planetary neighbor Mars? So far, the United States has not picked another objective beyond Viking, and budget pressures grow stronger every year. And the Soviets do not publish their plans and debates in a *Congressional Record*.

I consider the prime Mars mission for the next ten to fifteen years to be the automated return to Earth of a sample of Mars soil, similar to what was done on the Moon by the Soviets with Luna 16 (1969) and Luna 20 (1971). The return of a sample from Mars is far more difficult than it is from the Moon, but I feel that by the end of this decade the complicated entry and transportation technology associated with the mission will be within reach of either the United States or the Soviet Union. Which of us does it will be determined by desire as much as by technological capability.

I have a similar view about lunar exploration following Apollo. The United States has no present plans at all, while the Soviets apparently are continuing to develop even more sophisticated methods of unmanned exploration, utilizing roving Lunokhods and automatic sample-return devices. Perhaps other

THE LARGEST VOLCANO KNOWN TO MAN
This specially processed version of the mosaic of pictures of Nix Olympica illustrates its full size and detail. The summit caldera, which is about 40 miles across, is apparent in the center of the frame and sits at the top of a very large volcanic mountain. The sunlight is coming from the left. Around the base of the volcanic mountain is an escarpment of unexplained origin. The entire mountain is over 300 miles across.

surprises are still to come. I feel they will continue to work toward an ultimate goal, not just of a manned lunar landing, but a manned base in conjunction with an elaborate Earth-orbital space station. Again, this is the kind of enterprise that can be foreseen for about the 1980 time period.

In the case of Venus, the Soviets have been much more active than the United States, most recently carrying out the successful mission of Venera 8. I see no reason to suppose that their rate of activity will slow down—they have launched at every nineteenth-month Venus opportunity since 1960—and I would expect to see them deploy in the next two to four years the large "Proton" rocket system used for the Mars 2 and 3 missions.

The United States retains a unique capability for planetary discovery in the Mariner flight to Mercury by way of Venus to be launched late in 1973, the Pioneer 10 Probe, scheduled to take a first taste of the Jupiter environment in December 1973, and the Mariner/Jupiter/Saturn flybys planned for launch in 1977. These missions will provide first and lasting looks at the innermost and outermost portions of our solar system and, with Viking, will form the cutting edge of American scientific participation in space exploration through the balance of this decade. Robots will always precede man in his extension outward and will always provide the "first look" at new worlds. I hope the United States will continue to lead the world in this process. Perhaps more sophisticated remote sensing missions, such as orbiters of the planet Jupiter, or perhaps even Mercury, will follow in time, and perhaps some small entry probes into Venus also.

However, the main thrust of the American civilian space program for the next ten years is clearly aimed now at the development of new transportation technology—that is, the highly publicized Earth-orbital shuttle system, with less emphasis on new scientific achievements. It is my hope that the

HIGH-RESOLUTION DETAILS OF THE FLANK
A braided pattern of ridges and troughs, thought to represent lava fields, is shown in this high-resolution view of the flank of Nix Olympica. The long sinuous feature in the center of the photograph may be a collapsed lava tube standing along the top of a ridge.

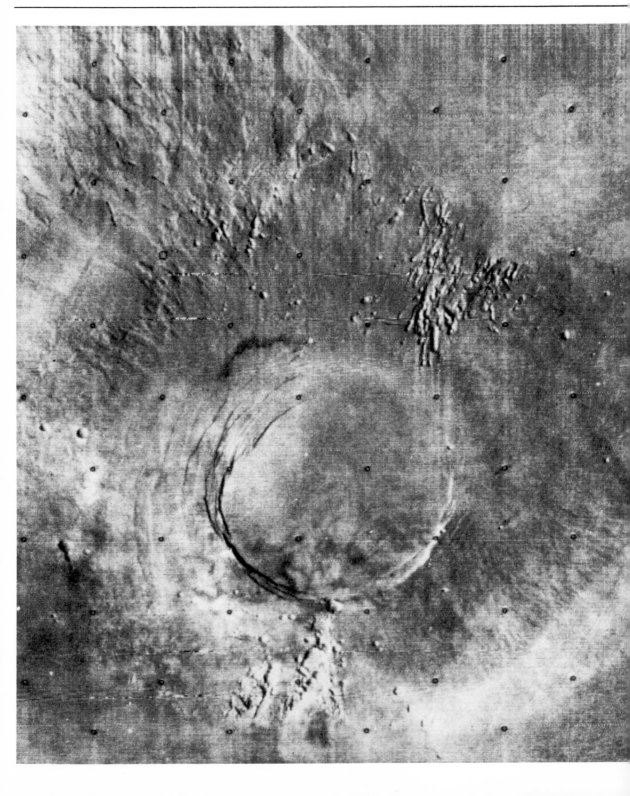

Viking landers and the Mariner flybys of Venus and Mercury in 1974 and later of Jupiter and Saturn will return sufficiently numerous and exciting pictures and other scientific information to keep the collective curiosity of this country stimulated. I hope that the excitement associated with Mars in the mind of man in the past, and through the present, can be extended to more distant and even more remote objects by the exploratory flybys.

It is interesting at this point to speculate why the United States and the Soviet Union now appear to be taking different paths in space after a ten-year period of rather similar objectives and approaches. The United States clearly is progressively reducing the priority of both manned and unmanned activities in space, whereas the Soviets are holding their own if not actually expanding their efforts. I can only interpret these national actions as genuinely representing differences in the priorities and attitudes of the peoples concerned.

The Soviet space enterprise, I think, primarily reflects domestic interests within the Soviet Union. Such foreign publicity as it provides is only of secondary importance in justifying the large expenditures involved. It seems to me inescapable, on looking at these two threads isolated within the total tapestry of this point in time, that the United States is turning inward a bit from what it could be and what it could do in space, while the Soviets are still enthusiastic over the prospect of leading the peoples of the world outward from their indigenous planetary captivity. Some may view the U.S. retreat with despair or even as indicative of decay within our civilization. But it can also be regarded as a sign of maturity in that we are at the same time attacking more fundamental problems of human existence than is the Soviet Union. We are, for example, noisily attempting to create a truly multiracial and multicultural society and to discover, for the individual, the family, and for larger groups, fundamentally new life styles that are more truly compatible

SOUTH SPOT
The southernmost dark spot seen in the photograph opposite page 47 also proved to be a large volcano capped by an enormous caldera, far larger than any others on Mars or Earth, as well as a series of ringlike troughs surrounding it, presumably arising from collapse.

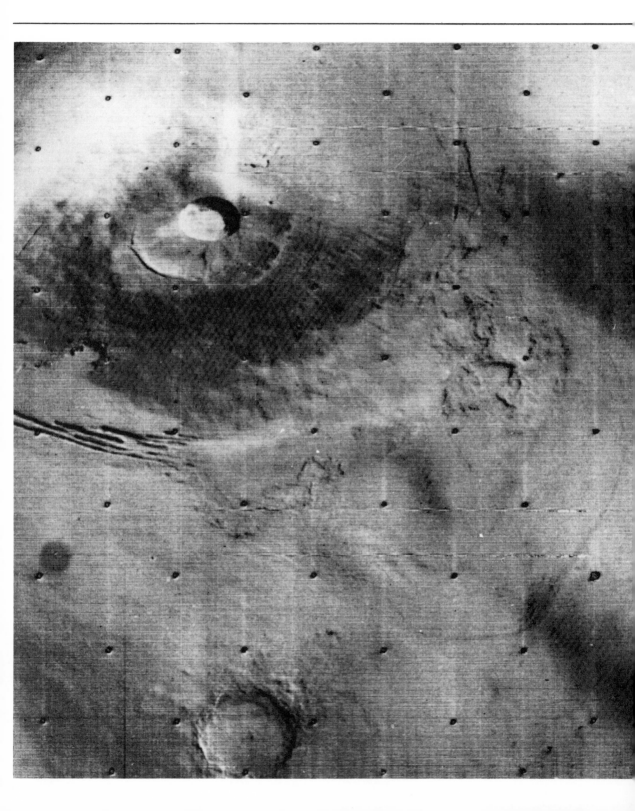

with the industrial revolution in which we are all imprisoned.

There is another way of assessing the future U.S. space program—from an economic and political point of view. The program inaugurated by President Kennedy with his May 1961 "An American to the Moon Before the End of the Decade" speech was created in a political climate totally different from the present one. First of all, the President and the Congress were of the same party, with substantial majorities, so that it was possible actually to create a national policy in space and implement it. Secondly, the major inflationary pressures that we have experienced so sharply in the last five to six years had not yet become so manifest. In fact, it was a significant element in the federal government's economic policy to stimulate various aspects of our economy through large federally sponsored technological endeavors. Finally, it was Kennedy's political genius to realize that the Apollo project not only appealed broadly to the imagination as well as the self-interest of a variety of political groups in the United States, but included a fixed date which was not negotiable, and thus did not permit the continued attrition and slippage that have afflicted so many other large technological endeavors.

The simple fact of the matter now is that the United States has not found a substitute for that Apollo approach to space. We've been unable to find a similarly compelling objective or to establish a more mature and sophisticated space policy to replace the rather simplistic but successful strategy of Apollo. Thus, what we have come to live with is a space program that does not reflect a real optimization of the whole but instead is an amalgam of those programs which are in the short term most defendable and most supportable in Congress as well as in the Administration. In addition, we are going into our fifth year of a divided Congress and Administration, which inhibits reorganization, redirection, and new momentum in space as well as in many other areas because of the lack of

MIDDLE SPOT
The middle spot of the three aligned dark spots in the photograph opposite page 47, like the other spots, is also a huge volcanic mountain, but this one is characterized by a perfectly circular single large caldera as well as radial scarps around the base.

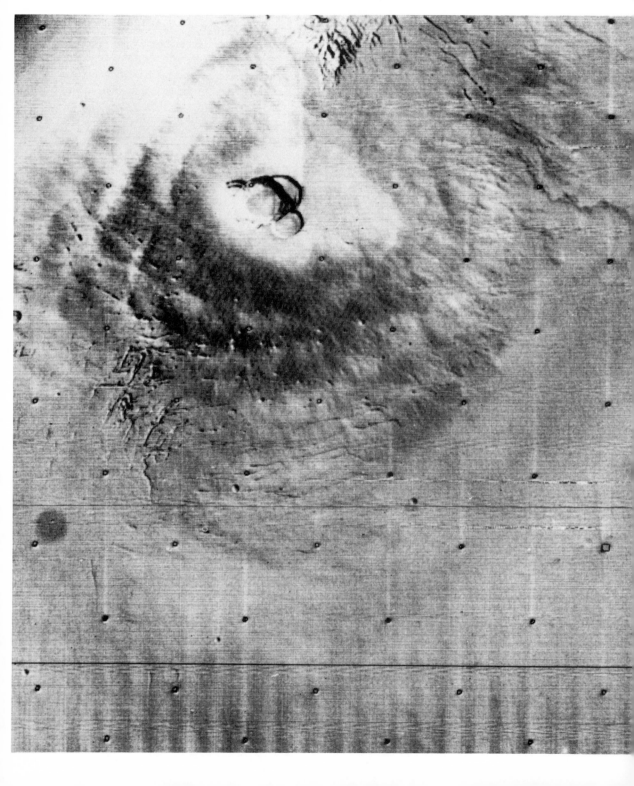

effective political leadership by any one segment of the political community.

Furthermore, the fundamental economic status of this country has changed radically since Kennedy's dramatic speech. There are real limitations to the use of the federal purse. It is no longer manifestly obvious that large technological programs are necessarily good and desirable *per se.* We face serious foreign trade deficits, which require more judicious use of both advanced technology and federal funds than were required in 1961. All these economic and political factors mean that we are living and will continue to live in a period in which it is much more difficult to establish bold new ventures in space, or perhaps in any other part of the American enterprise, than it was before.

It is entirely conceivable that the race for the Moon, exemplified by Apollo, may have been something of a historical anomaly similar to the race for the South Pole that took place shortly after the turn of the century and led to the gigantic triumph of Amundsen and the tragic death of Scott. Yet, at that point in time, the unimagined horrors of both World Wars still lay ahead. There was relatively little interest in further sustained exploration of the Antarctic areas until after World War II, when the International Geophysical Year took place. By then, transportation technology had become much, much cheaper. Antarctic exploration could be justified almost as an exercise to keep the Navy busy and did not require major development of new technology. Also, it was possible to carry it out under genuine international scientific collaboration, with some obvious practical benefits for the many nations participating.

Thus, it may be, if we could but view the tapestry of which we are a part from a vantage point of fifty or a hundred years hence, that the U.S. and Soviet race for the Moon and the surge in space technology of the 1960s were truly a bit anomalous. Perhaps in a decade or so transportation costs really will be

A CLEAR VIEW OF NORTH SPOT
A view of the northernmost spot seen earlier in the photograph opposite page 47 as it appeared after the clearing of the dust storm.

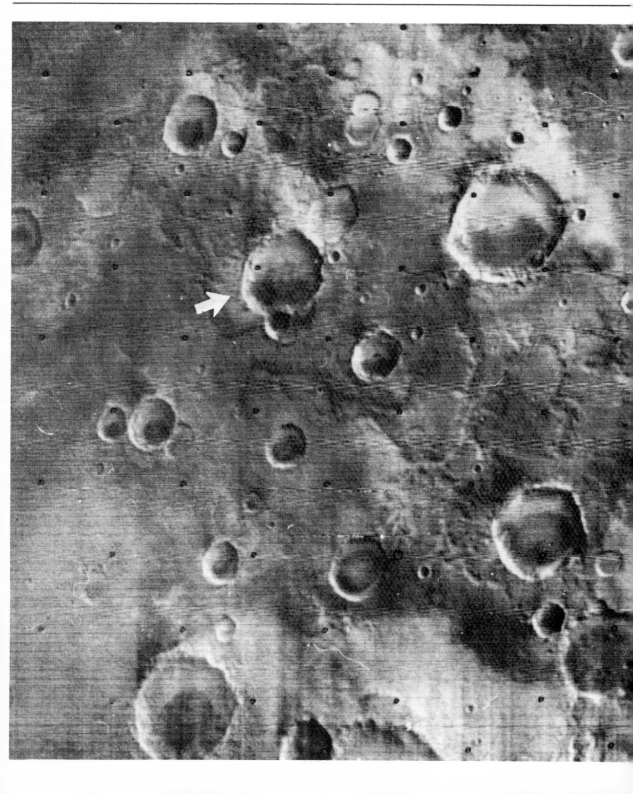

more tolerable and there will be a genuinely cooperative scientific exploration of the Moon, including renewed flights by American astronauts. A decade or two even further in time —if worldwide convulsions can be avoided—man's dominion may finally reach Mars. But what language will those first explorers speak? And will they care about the conscience and spirit of the individual?

In the case of the United States, the rapid Apollo build-up has been followed by a rather rapid dropoff, due to the volatile nature of our political system as well as, perhaps, to a rapid and painful maturing of our people. By contrast, the Soviets, being basically a technocratic* society dedicated to the ethic that technology can solve all their problems, still look upon space as the hallmark of the future and have not become sufficiently sophisticated as a people to appreciate the full complexities and limitations of that ethic. It seems to me that the United States is more advanced than the Soviet Union in concern over pollution and overindustrialization, for example. Most important, we have had our political revolution, with its resultant pluralistic, flexible—and volatile—governing system. It was a profound shock to me on visiting the Soviet Union to realize that, despite their revolutionary rhetoric, 1917 constituted primarily an economic change. Their rulers are just as dependent now on propaganda, censorship, and the secret police as were the czars. Technocracy will provide no help in meeting their eventual political crisis when the anguished cries of individual Russians finally mount to a deafening chorus demanding individual dignity, privacy, and civil rights.

Accompanying the winding down from Apollo, a new theme has developed through the U.S. civilian space effort—collaboration with the U.S.S.R. in space ventures. But each U.S. president

* Not to be confused with technocracy, the naïve, semifascist philosophy of the 1930s that proposed that technologists actually govern.

CRATERED TERRAIN
A view in which large craters dominate the scene. It was this kind of landscape, seen by Mariners 4, 6, and 7, which led to the idea that the surface of Mars strongly resembled that of the Moon. The crater indicated has been named "Airy."

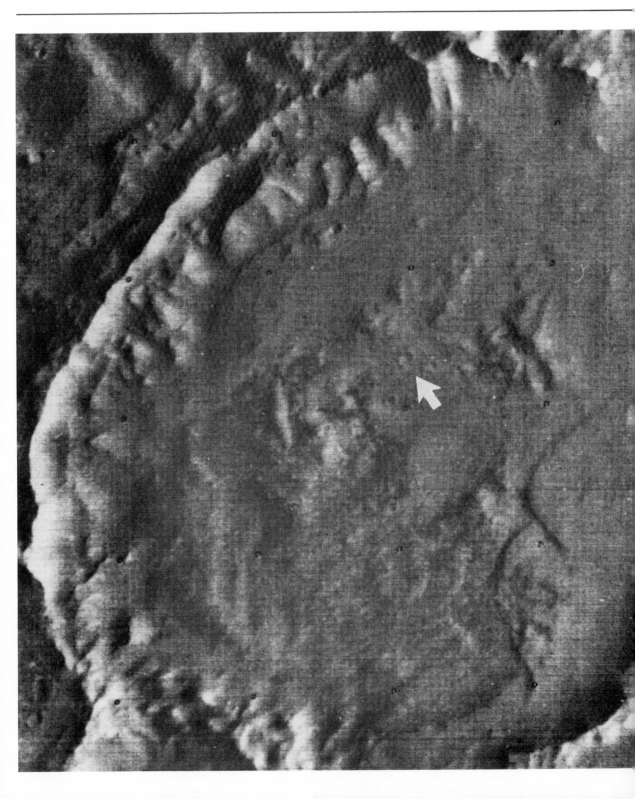

from Eisenhower to Kennedy through Johnson and finally Nixon has talked earnestly about the importance of international cooperation in space. Why did substantive cooperation between the United States and the Soviet Union begin only in 1971? Well, it takes two to collaborate, and if each is a superpower, then each must feel sufficiently confident in its national space achievements so that cooperation will not be regarded as a sign of weakness at home or abroad. I think that the first years of the 1970s included a historic approach toward parity in space (as well as strategic weapons) between the United States, with its Apollo Moon landing and still dominant position in planetary exploration, and the Soviet Union, with a growing and increasingly successful unmanned exploration program including automated return of lunar samples and the Lunokhod.

Thus, for a few years in the beginning of the decade of the 1970s there has been the delicate balance necessary for meaningful collaboration. As a result, a joint U.S.-U.S.S.R. Docking Maneuver of manned spaceships is scheduled for 1975. Each country has agreed to exchange "ground truth" from selected sites within its own territory to aid in cooperative environmental interpretation of low-resolution photography and other measurements from Earth orbit. The first steps toward significant collaboration in planetary exploration have been initiated. Both countries have exchanged lunar samples from different localities. More collaborative programs may emerge soon—in fact, about the only "new starts" in NASA programs these days seem to involve international collaboration in some way!

All of this space cooperation constitutes a dramatic contrast to the events of a decade ago, when out-and-out competition was the hallmark, just as SALT I and recent economic treaties, also based on valid cooperation of equals, constitute an encouraging change from the deadly stalemate foreshadowed by the Cuban Missile Crisis a decade ago. Thus, we may be enter-

CRATERED TERRAIN UP CLOSE
This high-resolution view of the crater "Airy" is of special interest in that it not only shows the details of such a crater but also includes the small crater "Airy-O," denoted with an arrow, which has been designated as the reference for the zero longitude on Mars—that is, equivalent to Greenwich, England, on Earth.

ing a truly different decade in space—and on Earth. Most manned space activities might well be part of joint programs. When American astronauts eventually return to the Moon, it may be for a tour of duty at a Soviet-U.S. base. The major unmanned missions to Mars and Venus in the eighties could conceivably be carried out jointly.

But the basic requirement for initiating cooperation—parity—is a necessary condition for continuing cooperation as well. Hence, if the era of collaboration in space exploration is to last through the next decade, the declining U.S. civilian space effort and the growing Soviet one will have to come to a stable equilibrium. Otherwise, space cooperation in the eighties would become an unequal partnership, a highly visible image of "second place" for the United States. Cooperative endeavors now provide a timely change in emphasis for the United States, when as a people we no longer feel the need to pay for unilateral space competition. But we shouldn't automatically assume the Soviets view it in the same light.

Certainly, Soviet scientists welcome it as another window to the West. But also there must be chauvinists in the Kremlin who feel driven to collaboration now as a result of our Apollo success, but view such an accommodation more as a temporary way station in a long-range technological competition in which there can be but one "first prize"—no joint awards. During the course of this decade we will learn from Soviet actions, especially concerning long-range technological developments in space, whether the current era of space collaboration is really the beginning of something new and hopeful, or just another change in scenery in the big power confrontation that has been the tragic hallmark of this century. If "2001" is to really come about, I feel it will have to be with international sponsorship. Rivalry in space is just a reflection of rivalry on Earth. Apollo and the Cuban Missile Crisis were both born of the

A RAW PICTURE
A picture of the surface of Mars as sent back by Mariner 9. This featureless print is what a typical Mars scene looked like before computer processing.

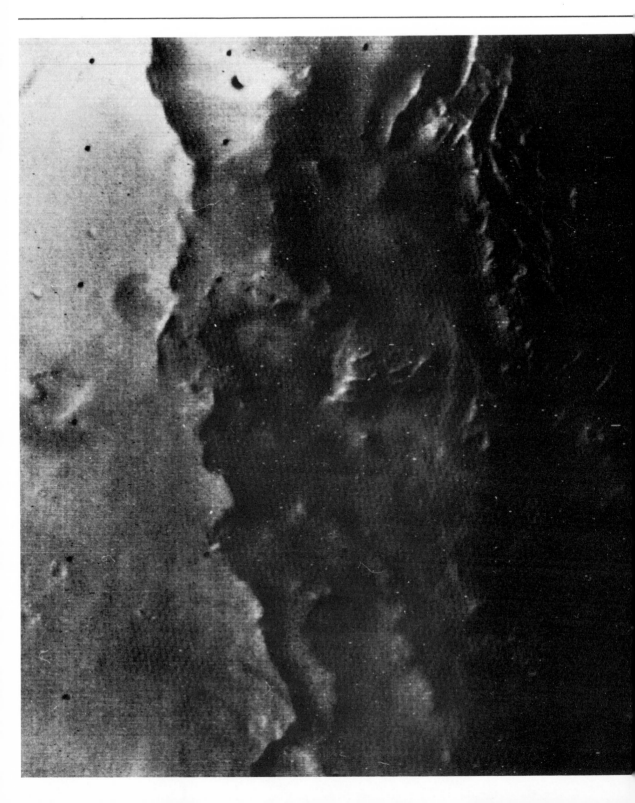

same causes. I don't think Earth can stand three decades more of that kind of rivalry.

These considerations of the economics, politics, and sociology of space, especially the U.S. space effort as compared to the Soviet, are perhaps worth pursuing to one final level. One can look on space as but one facet of what is really going on in history—the continued exponential explosion of the industrial revolution throughout all man's activities. The United States is the most advanced industrial country and is therefore at the cutting edge of this historic and shattering social and human experience. It seems to me that a major theme of the American experience in the past thirty or forty years, as it will certainly be in the next thirty or forty, is the extraordinary evolution of social and governmental institutions in response to technological change. A real question emerges in my mind concerning the rate of obsolescence of our institutions—universities, elementary and high school, churches, businesses, legislative bodies, professional societies, the National Academy of Sciences, and many others. All these institutions are evolving about as rapidly as is possible for the constituent individuals to evolve their attitudes about them. The high rate of suicide, insanity, and neurosis in our society is an index, I believe, of just that extraordinary rate of change we find ourselves immersed in.

This obsolescence of institutions and the agonizing rate of change about us can lead to despair. The present is not the key to the future, nor is it very similar to the past; so both conservatives and liberals are simultaneously disenchanted. This has been true for some time, and I suspect it will be true for some time into the future. I believe we are in the midst of the most rapid, and perhaps the most significant, evolution of man's social structure in his history. We have the mixed privilege of being the leading actors in that drama and are somewhat expendable in a historical sense. We have the op-

AN ENHANCED PICTURE
The same radioed data shown in the previous picture have been computer-corrected for the response of the vidicon tube and redisplayed with greater range relative to the photographic print, to produce the version seen above. The Soviet spacecraft Mars 3 returned some pictures during the dust storm also, but these were of no value for interpretation because they lack the tonal discrimination of the Mariner vidicon pictures and could not be enhanced on the ground.

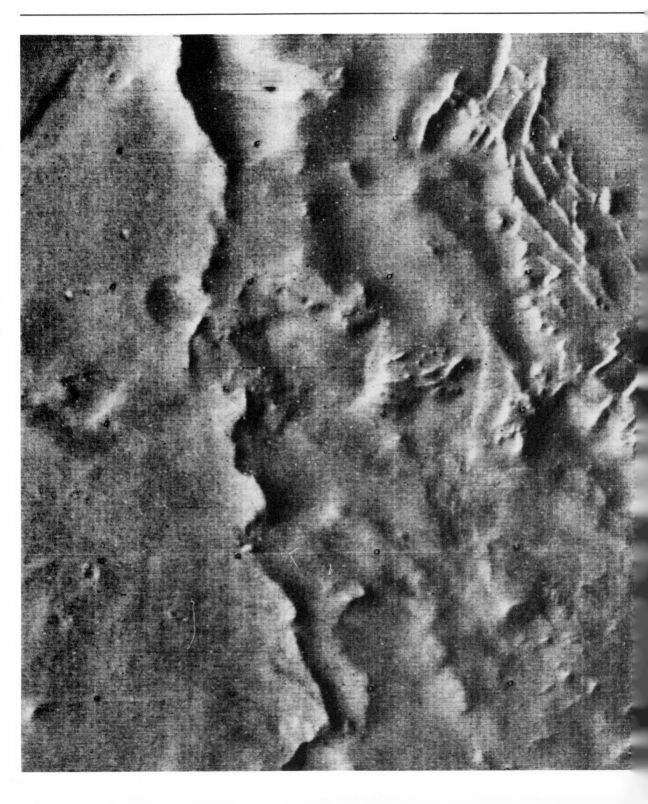

portunity at every step to lead the way and to create a society with achievements of lasting historical significance, including space exploration. During the last several hundred years, that, indeed, has been the American pattern. It could continue to be so for the next several hundred. We *can* continue at the forefront of the fateful evolution of man the toolmaker from his first halting steps in Africa several million years ago through the computerized, real-time, electronics world of today. It is likewise conceivable that the process of social and political change and the leadership which the United States has heretofore exemplified might be transferred to different peoples. Our identity could somehow become lost among other pressing themes shaping the context of the real year 2001, a year in which the color of men's skins on Earth will still loom larger, I feel, than even the luscious technicolor of Arthur Clarke's *2001*. It is up to us—all of us—to be what we *will* be, perhaps even what we *can* be. Space affords a kind of mirror of ourselves in which the character of our activity and the nobility of our goals are reflected back upon the terrestrial societies who would try their hand at it.

I do, in fact, maintain a genuine conviction of the Goodness of the historical process in which we are privileged to play a leading role, despite the accompanying horrors and dissatisfactions it engenders. I am hopeful that our nation can fulfill the promise of space that those of us in the forefront of this activity today can foresee. But I do not feel that we are required to in order for world history to be well served by present-day events.

On a more personal note, I find it difficult to maintain the level of optimism of a Bradbury or a Sagan. Yet, I cannot escape the observation that I must be basically a remarkable optimist nevertheless. My primary professional endeavor over the last four years has been getting ready for the Mariner flight to Venus by way of Mercury to be launched late in 1973.

A FILTERED VERSION
The data here have been processed to eliminate large-scale variations and brightness and to enhance those on a small scale, usually topographic detail. The scene shows a very unusual feature on Mars—a rectilinear pattern of what are probably ridges, standing up as though exposed by erosion. All three versions of the photograph were produced within five minutes of receipt on Earth by the Mission Test Computer system at JPL.

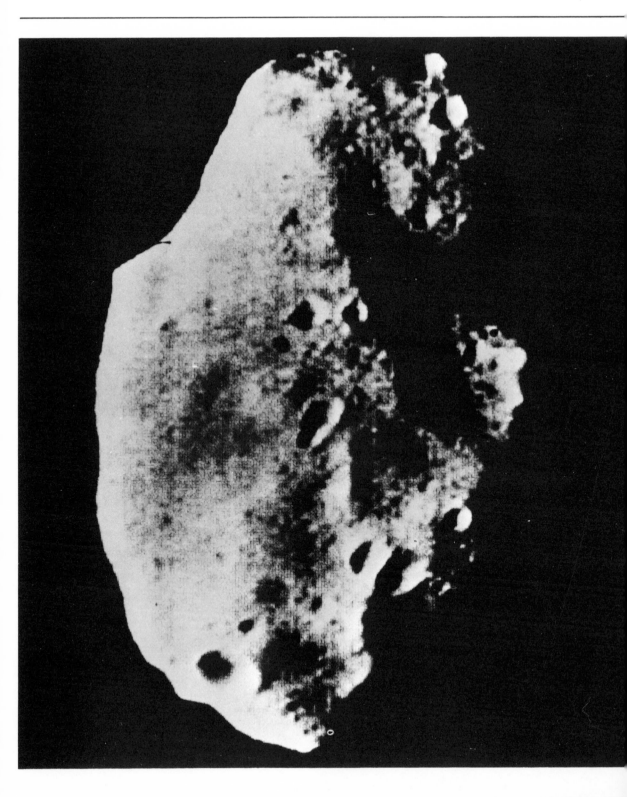

This activity has continued even during my intense participation with Mariner 9. What suggests a special degree of optimism is that but a single spacecraft (Mariner 10) will be launched on a single rocket—there is no backup, nor is there any plan for a subsequent endeavor in case of failure. In view of the fact that Mariner 1, Mariner 3, and Mariner 8 all lie at the bottom of the Atlantic Ocean, and that the Mariner 7 spacecraft was nearly destroyed by a battery explosion in flight, one cannot escape the conviction that anybody who commits a significant fraction of his life to a single launch of a single spacecraft, with no backup for a first look at Mercury, has basically got to be an optimist! Yet, if we are successful—despite the odds—a unique new thread will be woven into the great human tapestry, one that our parents could hardly have imagined and that our children's children will still remember.

PHOBOS
The larger of Mars' two satellites is shown in this computer-enhanced view produced by the Image Processing Laboratory of JPL. An enormous number of crater impacts make up its surface. The satellite is about 15 miles in diameter and is either left over from the original formation of Mars itself or is a refugee from the asteroid belt. Phobos is one of the darkest objects in the solar system.

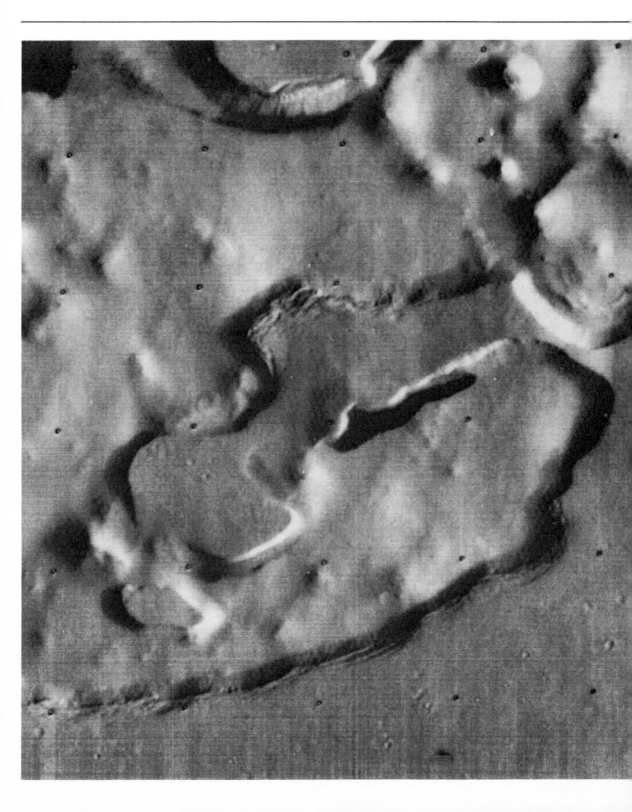

Arthur C. Clarke

Reading the transcript of our 1971 discussion is a curious experience, because it already seems to belong to another age—the prehistory of Martian studies. All of us knew that November evening in 1971, while Mariner 9 approached its moment of destiny, how important this mission might be, but I doubt that any of us would have dared to predict the full extent of its success. True, Mariner's cameras revealed no Martians carrying banners with the strange device BRADBURY WAS RIGHT (or even rival groups with NO—CLARKE WAS RIGHT). But what they did show was exciting enough, as the wonderful photographs in this book amply prove. At last, we are zeroing in onto the *real* Mars.

For most of this century Mars has been haunted by the ghost of Percival Lowell, the man with the tessellated eyeballs. Mariners 4, 6, and 7 started to exorcise that ghost; Mariner 9 completed the job. The famous "canals" are gone forever. Why they ever appeared in the first place could be material for a valuable study of psychology and physiological optics—and incidentally, the time is now more than ripe for a modern biography of Lowell, surely one of the most fascinating characters in the history of astronomy.

Now that we have good-quality photographs of Mars, someone should compare Lowell's drawings with the reality to try and find just what happened up there at Flagstaff at the turn of the century. How was it possible for a man to sustain a self-consistent and extremely detailed optical illusion (if that *is* what it was) over a period of more than twenty years? How did he convince others of his vision? What correlation, if any, was there between the ability of other astronomers to see the canals, and their position on the Lowell Observatory payroll? These are just a few of the questions that might be asked. . . .

Recent work on the nature of vision has shown that the eye

MESAS
Sharp cliffs form the boundary between a smooth plateau, which exhibits a few small impact craters, and an eroded area exhibiting hummocky terrain. One part of the plateau is entirely detached in the upper right-hand side, to form a mesa adjacent to a large tongue stretching out toward it running from the lower left to upper right.

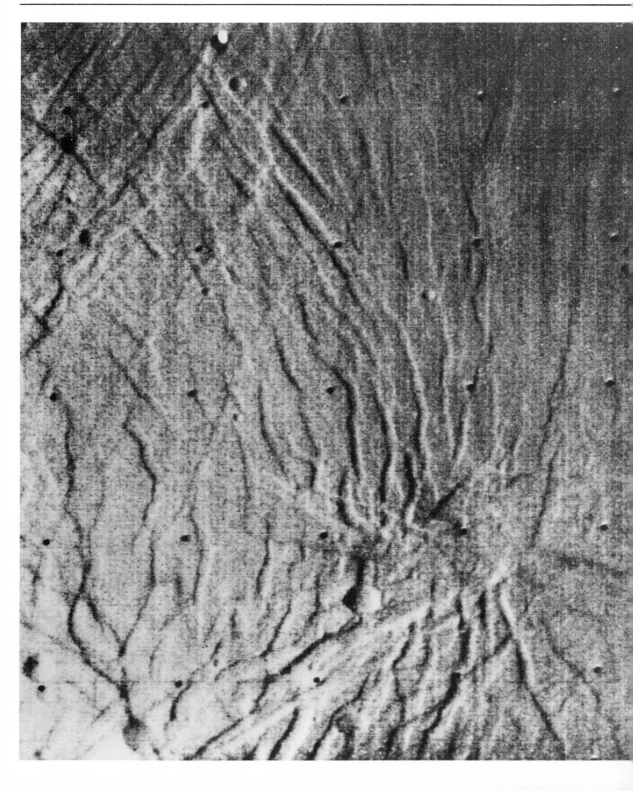

is capable of feats which, *a priori,* one would have said were completely impossible. "Eidetic imagery" is an example which may be very relevant here. Dr. Bela Julesz of Bell Labs discusses a case* where a subject was able to store an *apparently random* pattern of 10,000 picture elements—a 100 x 100 matrix of dots—and fuse it *twenty-four hours later* with another array to produce a stereoscopic image! As Bela Julesz remarks with considerable understatement, "These experiments appear incredible," but they do suggest that the eye-brain system has an astonishing capacity to store detailed images. Could Lowell have built up over a period of years a largely mental picture of Mars from the fleeting patterns glimpsed through his telescope? The mind has an extraordinary ability to "see" things that are hoped for, assembling any chance visual clues that may come to hand—to use a slightly mixed metaphor. When you are expecting to meet a friend in a crowd, how often you see him before he really appears!

If Lowell's Mars was indeed a largely subjective one, it also had to be dynamic. It must have changed continually with rotation, distance, seasons, to match the changing appearance of the real Mars. Certainly a fantastic feat of creative imagination, of the greatest interest to psychologists. . . .

And although I'm now speculating in areas with which I'm completely unfamiliar, I'd like to stick my neck out just a little further. Could there be some connection between Lowell's superbly maintained and brilliantly proselytized delusion (remember, plenty of other observers "saw" the canals) and a similar phenomenon of our own time? I don't think there can now be any doubt that hundreds of intelligent, sober, and altogether reliable citizens have honestly "seen" brightly moving lights in the sky and all the other familiar UFO phenomena.†

* *Foundations of Cyclopean Perception* (Chicago: University of Chicago Press, 1971).
† Anyone who still doubts this should read Allen Hynek's *The UFO Experience* (Chicago: Regnery, 1972).

CRACK PATTERNS
Several different directions of cracks 1 to 2 miles across and many tens of miles in length make up the surface of a high plateau.

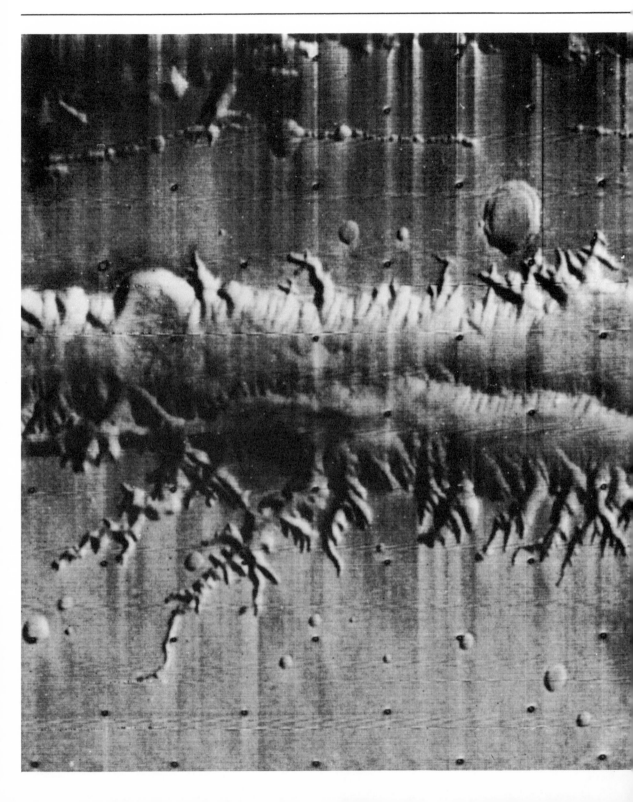

How many of these originated in some such process as Lowell's discovery of canals?

However, back to the real Mars. It now appears that, by one of those ironies not uncommon in science, the earlier Mariner results caused the pendulum to swing too far to the other extreme—away from the hopelessly romantic view of Mars. For the few years from 1965 to 1972 Mars was a cosmic fossil like the Moon—no, not even a fossil, because it could never have known life. The depressing image of a cratered, desiccated wilderness was about as far removed from the Lowell-Burroughs fantasy as it was possible to get.

There were some, undoubtedly, who accepted the new "revelation" with considerable relief—even glee. Now there would be no further fear of that dreaded cry in the night: "The Martians are coming! The Martians are coming!" We were comfortably alone in the solar system, if not the universe. . . .

Well, perhaps we are, but it seems more and more unlikely. The new Mars that has suddenly emerged from the Mariner 9 photos, a world of immense canyons and volcanoes and erosion patterns and—dare one say?—dried-up seabeds is a much more active and exciting place than we would have ventured to hope only a few years ago. Lowell and Company may yet have been partly right for the wrong reasons.

It is not really a coincidence that while Mariner 9 was being built, the first positive evidence for the chemical evolution of complex organic molecules beyond the Earth was being discovered. The basic building blocks of life were being found in *meteorites,* of all places—perhaps as hostile an environment as could be imagined. In view of this, and the obvious signs of past water activity shown in the Mariner photos, the biologists will have some explaining to do—if there is *no* life on Mars.

Meanwhile, we science fiction writers had best be cautious for a few years—perhaps until U.S. or U.S.S.R. soft-landers

THE HUGE CANYON OF MARS
This view shows about 300 miles of the very extensive east-west canyon system that prevails in the equatorial region of Mars. Branching tributaries in the bottom of the picture apparently have been the site of landslides and mass movement from the higher, sparsely cratered plateau into the canyon bottom itself. The canyon is about 2 to 3 miles deep in many places.

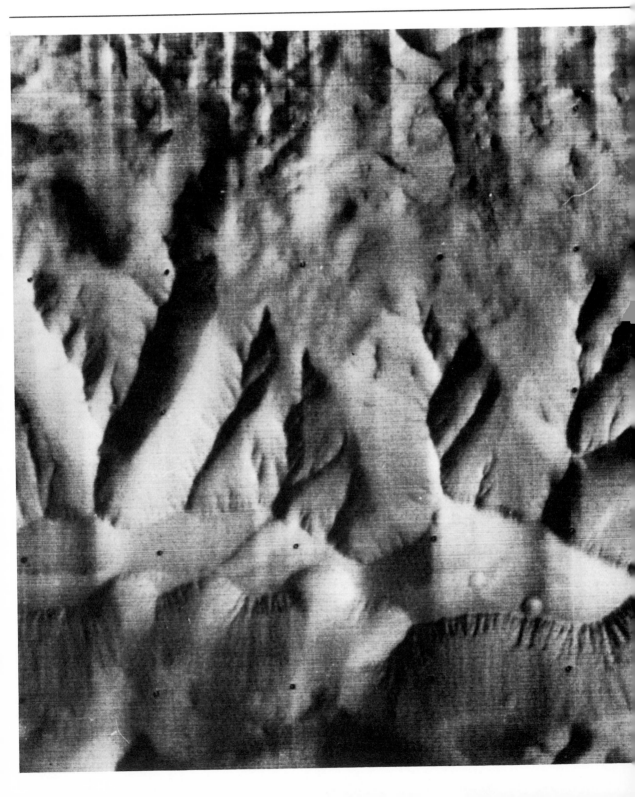

start to do some detailed reporting in the mid-seventies. For myself, I'm already a little embarrassed to see that *The Sands of Mars* (1951) contains the sentence, *"There are no mountains on Mars"* (in italics, yet . . .). Well, it took over twenty years to shoot *that* one down, so it had a good run for its money. And on the plus side, we now have some perfectly beautiful photos of Martian sand dunes (page 92), so at least my title was completely valid. The sands of Mars have survived very much better than the oceans of Venus. (Poor Venus —what a hatchet job the Mariners and Veneras did on *her!* But that's another story.)

There are some not-very-bright and/or badly educated people who complain, with apparent sincerity, that scientific research destroys the wonder and magic of nature. One can imagine the indignant reaction of such poets as Tennyson or Shelley to this nonsense, and surely it is better to know the truth than to dabble in delusions, however charming they may be. Almost invariably, the truth turns out to be far more strange and wonderful than the wildest fantasy. The great J. B. S. Haldane put it very well when he said: "The universe is not only queerer than we imagine—it is queerer than we *can* imagine."

I feel sure that Mariner 9—and its successors—will provide many further proofs of this statement. We have already learned an instructive lesson from the Moon, which is becoming more complicated and more interesting with every expedition. The same thing will happen with Mars.

Whether we find life or not, we will discover things which we could *never* have imagined. And these will provide material for the deeper and richer fantasies of the future, just as the earlier observations inspired the fantasies of the past.

And the beauty of it is—we can have it both ways! When men are actually living on Mars, at the turn of the century, they will be reading the latest works of the lucky science fiction

CANYON DETAIL
Rough terrain on the walls and floor of the canyon shows what are probably the effects of avalanches and other kinds of mass movements. A remnant of the original uneroded surface, seen in the lower portion of the picture, is capped by a small impact crater. The area shown is about 15 miles wide.

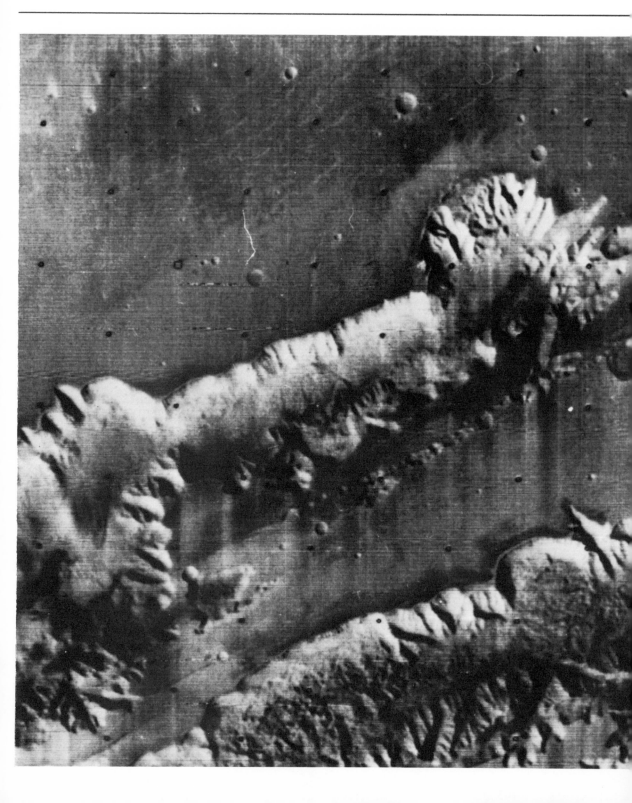

writers who are starting their careers now, at the beginning of the Fourth Golden Age. Yet at the same time, they will still be able to enjoy, from their new perspective, the best of Wells and Burroughs.

And, I hope, of Bradbury and Clarke. . . .

TWIN CANYONS

Two parallel canyons are separated by a remnant of a plateau with a crater chain along it. The area shown is about 300 miles wide.

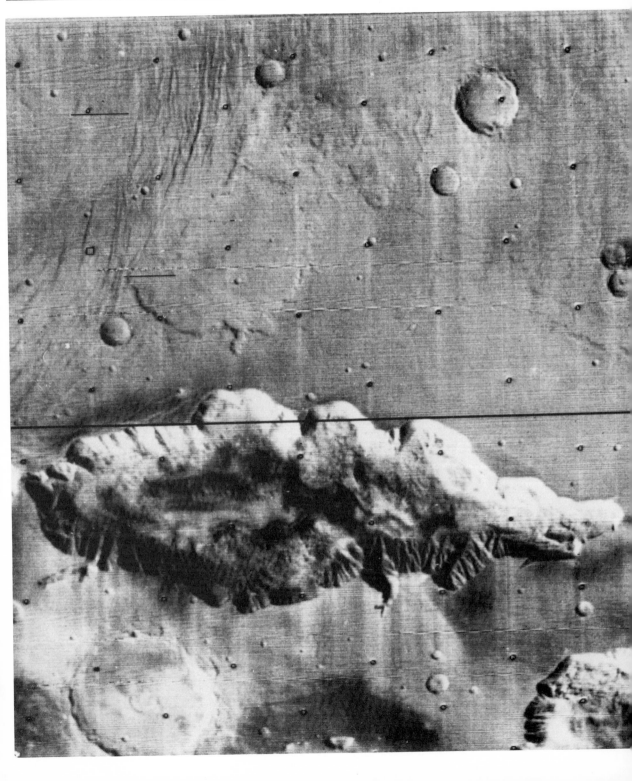

Carl Sagan

This book spans a fundamental transition in our knowledge about the planet Mars. It begins on the eve of the injection of Mariner 9 into Mars orbit—at a time, so clearly delimited in the first part of this book, characterized by poor data, wishful thinking, overcautious conservatism, a strange kind of Earth-Moon parochialism, and too sweeping generalizations from a few good facts to views of the whole of Mars. Now we have moved from a data-poor, theory-rich situation to one that is data-rich, theory-poor. Mariner 9 is the first spacecraft of mankind to orbit another planet. We are inundated with hard facts. The television cameras alone obtained more than 7,500 photographs of the planet, mapping the entire surface to a resolution of one kilometer, and a small percentage of the surface to a resolution of about 100 meters. We have thousands of spectra—ultraviolet spectra, giving information on the surface topography, atmospheric aerosols and composition, and the temperature of the upper atmosphere from which leakage of molecules to space occurs; and infrared spectra, providing data on surface topography and composition, atmospheric structure and winds, and clues to minor constituents. The surface has been peppered with infrared radiometer examinations of the temperature variation through the day, giving insights into the surface thermal properties and porosity. More than a hundred places on Mars have been examined by the S-band occultation experiment, giving the structure of the atmosphere and ionosphere above these places, and the distance of these places from the center of Mars. And the celestial mechanics experiment has begun to map the distribution of mass in the interior of the planet.

The Mars revealed by Mariner 9 corresponds to few comprehensive views of the planet imagined earlier. There are certainly no canals as Schiaparelli or Lowell drew them.

89

GIANT CLOSED CANYON
In this part of the great canyon system, a huge area of the canyon, more than 200 miles long and 2 to 3 miles deep, is entirely closed off, demonstrating that continuous drainage is not necessarily a property of the canyon system of Mars. The surrounding territory is a sparsely cratered, more elevated area.

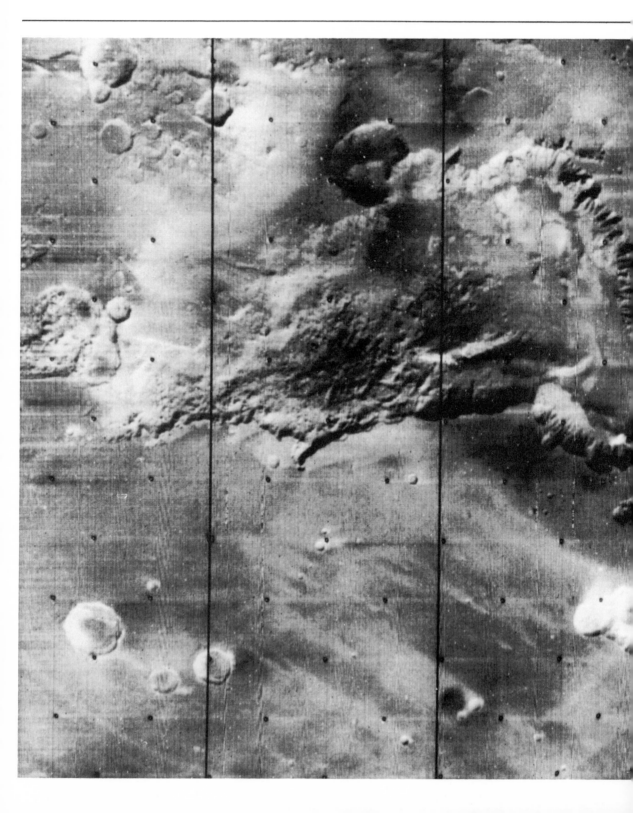

Mariner 9 has examined enough of Mars at high enough resolution to exclude the presence of a civilization of terrestrial extent and level of development. Not only are there no signs saying "Bradbury was right," there are no artifacts of any sort visible with 100-meter resolution. The planet-wide feudal-technological civilization envisioned by Edgar Rice Burroughs does not exist.

But neither is Mars like the Moon. There are cratered terrains, true, but there are large regions on Mars breathtakingly different from our natural satellite. Enormous volcanoes rise ten to twenty miles above their surroundings. Apart from the polar caps, they were the first features on the planet seen despite a severe dust storm when Mariner 9 went into orbit around Mars in mid-November 1972. The peaks of the volcanoes were sticking up through the dust. As the storm cleared we gradually obtained better views of the flanks and calderas of these volcanoes. They are enormous features. The biggest, Nix Olympica, is larger than the largest such construct—the Hawaiian Islands—on the planet Earth. The flanks of these volcanoes are remarkably free of impact cratering, implying that they have been produced in geologically recent times, perhaps only in the last tens or few hundreds of millions of years. This means that Mars is, today, geologically active—quite a different picture from that of a geologically inactive Moon, which was popular only a few years ago. Indeed, in the interim between the beginning and the end of this book, even the Moon has begun to look less like the Moon that was once thought to be. Apollo 16 seismic observations and Apollo 17 surface geological investigations have given some indication that the Moon also may be geologically active, at least on a small scale.

But Mars is active on an enormous, colossal scale. Not only are there the huge volcanoes, there is an array of linear features —not like canals, generally not in the positions of the old

CHAOTIC TERRAIN

This area of jumbled and depressed terrain was first discovered by Mariners 6 and 7 in 1969 and then more extensively mapped by Mariner 9. It is a site of unusual erosional forms, perhaps related to withdrawal, which is growing at the expense of the moderately cratered terrain surrounding it. The area shown is about 250 miles wide.

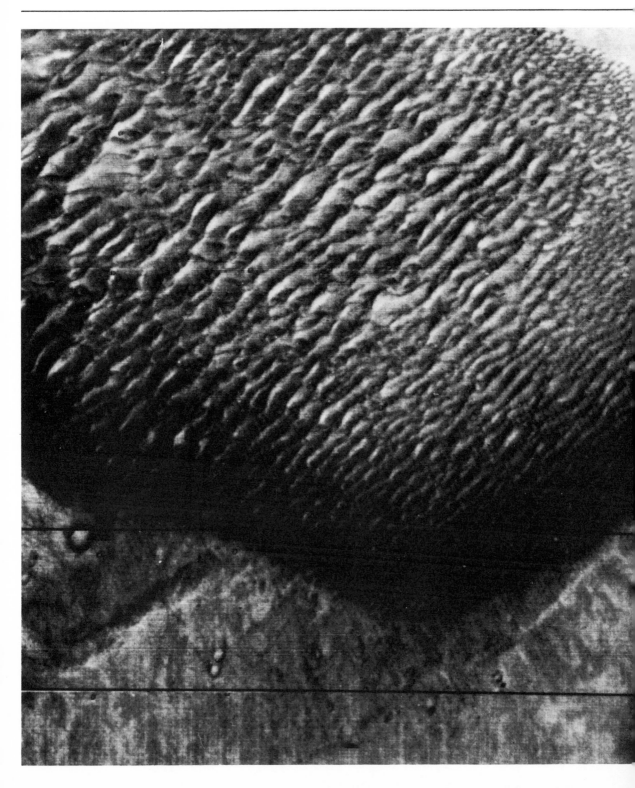

canals, and generally not visible in any case from Earth, but nevertheless approximately linear gouges, stretch marks on the skin of Mars. The greatest of these, the enormous Coprates Rift Valley, runs east-west for 80 degrees of Martian longitude and is comparable in extent to the great East African rift valley system, the largest on the planet Earth. We do not know if the Coprates valley was produced, as the East African rift valley was, by continental drift, a sign of convection in the interior of the planet and of great geological activity. But whatever the origin of these gouges, they also speak eloquently for a geologically vigorous Mars.

As the dust storm on Mars cleared I was astonished and delighted to see, layered down on the Martian surface for our edification and appreciation, a set of natural weather vanes and anemometers. Thousands of Martian craters have bright or dark tails of streamlined shape neatly emanating from them. In a given region most of these tails are parallel. We think they are caused by dust trapped in the crater during a dust storm and blown out by a prevailing wind in the final stages of the storm. They need correspond to only a thin veneer of dust, brighter or darker than the surroundings—perhaps only a few millimeters thick—but they trace out for us the directions of the high winds. In the Martian tropical zone, these crater tails show a clear tendency to follow the prevailing winds on Mars, calculated from meteorological theory and observed indirectly by the infrared spectrometer on Mariner 9. At higher latitudes, the winds caused by the general circulation of Mars—produced by the unequal heating of equator and pole—are predicted to be weak. The streaks we see in these regions are due to other sorts of winds: winds driven by the enormous elevation differences on Mars (we see some sign of winds running downhill from the great volcanic plains); winds like those dust devils in the American southwest; and, near the polar cap, winds driven by the great temperature

SAND DUNES
This high-resolution view of a dark region in the center of a large crater shows what very probably are great sand-dune fields somewhat similar to those that form in arid areas of the Earth. The area pictured is about 30 miles wide.

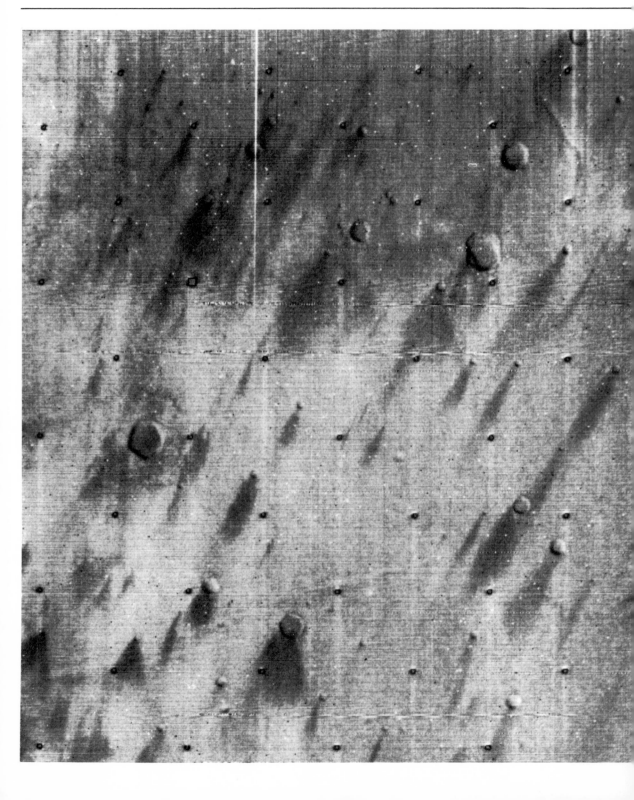

difference in summer between adjacent frosted and unfrosted terrain. In some places the wind streaks show several directions, probably corresponding to the activity of high-velocity winds at several different times.

One striking result is the discovery that the outline of dark streaks on the Martian surface corresponds quite closely to the outlines of the dark markings viewed during a century of groundbased observations. What is more, the locales of Mars known by groundbased observations either to vary regularly with the seasons, or to vary erratically, correspond to the areas of Mars in which changes are observed in the constituent wind-blown streaks.

The seasonal changes on Mars have been attributed, at least since the time of Lowell, to vegetation responding to the warmth and wetness of the Martian spring. Lowell even proposed that we were seeing cultivated crops in these annual contrast changes. The Mariner 9 observations seem to have confirmed that they are due instead to variations in the patterns of wind-blown dust, deposited and lifted by the seasonally varying high-speed winds on Mars. Thus, the seasonal changes appear to be due to meteorology rather than to biology. At the same time, there is nothing in these observations that excludes biology, and indeed, at times of great dust storms ultraviolet light at the surface is significantly attenuated, and if microorganisms did exist they could be rapidly dispersed over the whole of the planet.

There are other clear signs of windblown dust on Mars. Mariner 9 found that the insides of many craters have a dark patch or splotch. We find that this splotch appears very frequently on the part of a crater interior that corresponds to the direction from which the streaks from this or adjacent craters emanate. A crater with a dark splotch and bright streak, we think, is most likely produced by winds lifting bright particles from the interior of the crater, thereby revealing to view the

DARK TAILS
These dark swaths tailing out from individual craters constitute a striking pattern and are believed to be associated with the large-scale markings observed from Earth through telescopes. The lighting is from the left, and the view is about 300 miles wide.

underlying dark material, and then depositing these bright particles outside the crater and downwind.

Such splotches require the transport of only thin layers of dust, but other splotches are revealed as enormous sand-dune fields. The best developed of these bears a remarkable similarity to the Great Sand Dunes National Monument in Colorado. Here we see clear evidence of the long-term effects of prevailing winds on mobile particulate matter. By photographing the same region successively in the course of the mission, we have uncovered many instances of dark streaks and splotches that are slowly increasing in size. We have seen the dust-transportation in progress.

Because the Martian atmosphere is so thin, higher winds are necessary to make sand particles move than on the Earth. I believe that the minimum wind speed in the middle of the atmosphere necessary to make a dust grain roll over on the surface is 50 to 70 meters per second (110 to 150 miles per hour), compared to only a few meters per second for sand movement on the Earth. Therefore the erosion and abrasion due to windblown sand on Mars will be very great. I calculate that in places of strong prevailing winds, the abrasion rate may be as high as a tenth of an inch to an inch per year. So far as I know, none of the previous accounts of Mars—fictional, mystical, or scientific—has had a word to say about this pervasive aspect of the Martian environment. The high winds and mobile dust, in addition to causing the bright and dark markings and the seasonal changes, pose a significant hazard for space vehicles landing on Mars. Indeed, it is not unlikely that the failure of the Soviet Mars 3 entry probe in December 1971 was connected with the high winds during the dust storm.

Perhaps the least expected finding from Mariner 9 is that Mars appears to be covered with a great variety of irregular channels—some of which have meanders and tributaries (and tributaries of tributaries) and do not start or end in a crater.

BRIGHT TAILS

In this view, bright tails emanate out from the craters rather than dark ones, thereby complicating any simple interpretation of this kind of marking. The area shown is about 30 miles wide.

We find that these sinuous channels are strongly concentrated near the Martian equator—a fact which points directly to the conclusion that the channels depend on temperatures higher than normal for Mars. We have a situation where features have been carved out by a liquid flowing steadily on the Martian surface. If this liquid is to be made of not exceptionally exotic stuff, and requires high Martian temperatures for its flow, it can only be water. But liquid water cannot exist on Mars—or at least the Mars we see today. The total pressure is not enough to keep water in a liquid state for any extensive period of time. I am therefore led to the conclusion that these channels were carved out at a time when the Martian environment was significantly different from what it is today—a time of higher pressure, higher temperature, and higher water abundance. Because the channels are comparatively fresh, this epoch could not have been in the earliest stages of Martian history.

From the ultraviolet spectrometer experiment on Mariner 9 we know that substantial quantities of water could not have escaped from Mars even in the entire course of its existence. Therefore, if the channels are truly river basins, the water that carved them out must still be on the planet today. If Martian organisms exist they may find the deserts of Mars more like oceans. I can very well imagine a kind of Martian organism that drinks the water of the rocks of Mars. And there is undoubtedly subsurface permafrost—water-ice frozen at low temperatures beneath the Martian surface.

But the great repositories of volatile gases on Mars are the polar caps. Mariner 9 has indicated that the total thickness of the permanent polar caps of Mars is about a mile. If somehow all this frost were converted into gas, it would result in a total pressure all over the Martian surface of about one atmosphere —that is, about the same as the surface pressure on the planet Earth today. Moreover, the close-up photographs of the permanent polar caps show an alternate bright and dark

BLACK SPLOTCHES
This striking view of dark markings surrounding a small crater shows the remarkable diversity of forms in the markings on Mars. This area also exhibited a change in markings during the Mariner 9 mission. It is about 50 miles wide.

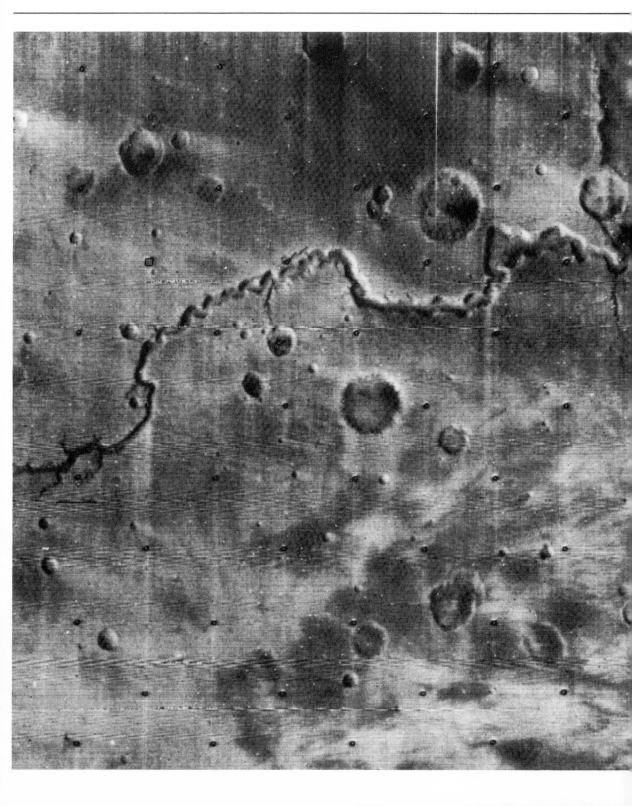

banding or laminae, which may be caused by alternating epochs depositing ice and dust, ice and dust. Thus, both the channels, which presumably can be produced only in a very different environment from the present Martian environment, and the laminae point to a major variation in the Martian climate.

How could such a variation occur? Perhaps something like this:

We begin with Mars as it is today, in a vast global ice age, with a large frozen atmosphere at the polar cap. Then we assume an epoch in which dark dust is deposited at the polar cap, perhaps by events like the great dust storm. Dust deposited at the poles causes a greater absorption of sunlight there, resulting in a slightly greater atmospheric pressure. The heat transported by the circulation of the atmosphere from equator to pole now goes into heating the polar cap somewhat more efficiently. The atmospheric pressure now becomes even greater; hot air transported from equator to pole is even more efficient in heating the pole; and we have what I call an advection runaway, which continues until much of the material in the polar cap is vaporized. As a result of the runaway the Martian climate has moved entrancingly towards a much more Earthlike climate, and it is at these times, I propose, that the Martian channels are carved out by running water near the equator.

The development of today's conditions occurs with all the factors running the other way: a long period of no dust at the poles, with the result that they remain bright and thus absorb less sunlight, get colder, and cause more frost to deposit there. The atmospheric pressure then declines, heat transport from equator to poles is less efficient, the poles cool even more, and we reach a situation more or less like the present one.

While I see no significant impediment to there being con-

SINUOUS VALLEY

This striking sinuous valley, extending 300 miles on the surface of Mars, resembles in some ways a dried-up river bed on Earth. However, features on the Moon such as Hadley Rille also exhibit such sinuosity. The branching tributaries seen at the lower left-hand side are reminiscent of those along the great canyon. Features like this have led to the speculation by some of the possibility of water on the planet in the past.

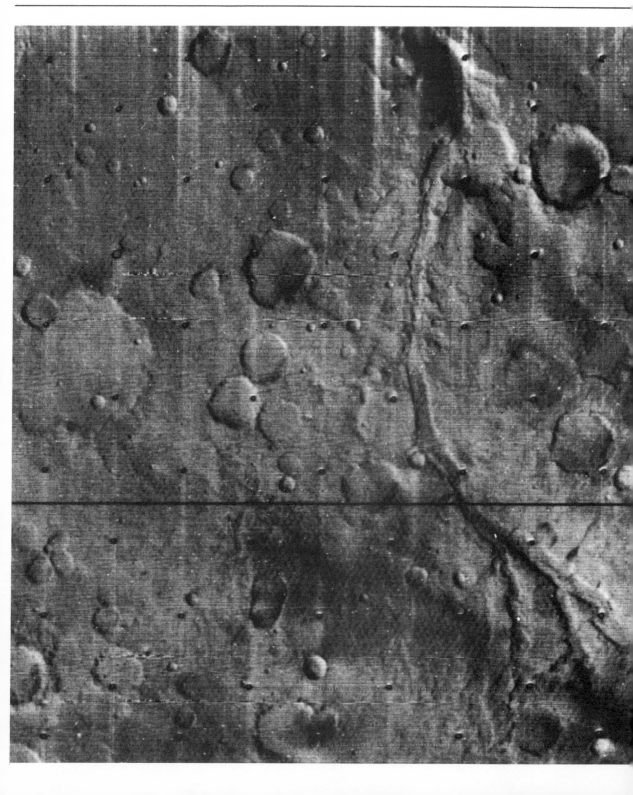

temporary biology on Mars, it is obviously easier to imagine a biology on Mars at the time of more Earthlike conditions. And it is not out of the question that there are organisms on Mars in hibernation or other sorts of biological repose, awaiting the end of the Martian ice age.

In *The Sands of Mars,* Arthur Clarke imagined a long-term biological reconditioning of Mars—making it more habitable for human beings by the appropriate breeding of plants from Earth. The foregoing ideas on climatic variation on Mars suggest that much more Earthlike conditions could be brought about periodically and that we might hasten the return of more clement conditions by adjusting the amount of dark material on the polar caps. Such human intervention into the Martian environment should only be undertaken, if at all, after a long-term, comprehensive program of studying the present Martian environment, both physical and biological.

I suppose it is just barely possible that the Mars that Lowell and Burroughs and Bradbury have written about may have existed in the past, or will exist in the future. But I wouldn't bet on it. If there is life on Mars, it will probably turn out to be astonishingly unlike any sort of life on Earth—unless we muck up the planet by failing to sterilize our spacecraft.

The data provided by Mariner 9 suggest that, at least at some times and in some places, Mars may be much more habitable for terrestrial microorganisms than many had suspected possible. They also show us dramatically that ultraviolet light from the Sun (which can fry terrestrial microorganisms in approximately one second) can be prevented from striking the surface by dust in the atmosphere, and that particles the size of microorganisms can be rapidly transported all over the surface of the planet. These factors, taken together, make it even more urgent to insist on strict sterilization of spacecraft intended for Mars landings.

At the meeting of COSPAR (the Committee on Space Re-

ANOTHER SINUOUS VALLEY
This view of a sinuous channel with branching tributaries is part of a 450-mile-long feature on the Martian surface believed by some to have been formed by flowing water. The remaining portion is shown in the following photograph. Other scientists think unusual kinds of volcanic activity might have been involved, and still others regard such channels as unexplained features.

search) held in Madrid in May 1972, Professor V. I. Vashkov of the Soviet Ministry of Health outlined in detail the procedures used to sterilize the Mars 2 and 3 entry vehicles—the first space-craft from Earth to land on the surface of Mars. Vashkov told of an elaborate and exceptionally careful protocol involving heat, gaseous sterilization, and high-energy radiation.

The United States Viking Program has similarly elaborate plans—involving heat soaking of the entire spacecraft and subsequently enclosing it in a shroud—to prevent its carrying terrestrial microorganisms to Mars. The dangers in con-tamination of Mars are (1) the possibility of landing organisms which will then be detected by our life detectors—surely an extremely expensive way to investigate common terrestrial microorganisms—and (2) the possibility of doing ecological damage to a Martian biota, if it exists. It is very rewarding to see, despite the cost of sterilization, so responsible an attitude taken by the two great spacefaring nations.

There was one further bonus from Mariner 9—the first close-up photographs of Phobos and Deimos, the Martian satellites, which had been thought by some to be artificial satellites launched by an ancient Martian civilization of very great powers. Instead we find both satellites to be old, dark, battered, and entirely natural objects. This will probably come as a dis-appointment to some. The two possible explanations of the origins of Phobos and Deimos are, however, almost as interest-ing: either they are captured asteroids, in which case we have had our first close-up look at these elusive inhabitants of our solar system; or, as suggested by Alfred Russel Wallace, they are debris left over from the origin of Mars.

Alfred Russel Wallace continues to impress me—partic-ularly in light of his time and scientific background—as the man who came closest to guessing or deducing what the real Mars is like. Another man who comes to mind is not often mentioned: the late Dean McLaughlin, a professor of Astronomy

SINUOUS VALLEY, CONTINUED
A continuation of the valley seen in the preceding photograph. Lighting is from the left in both of these photographs.

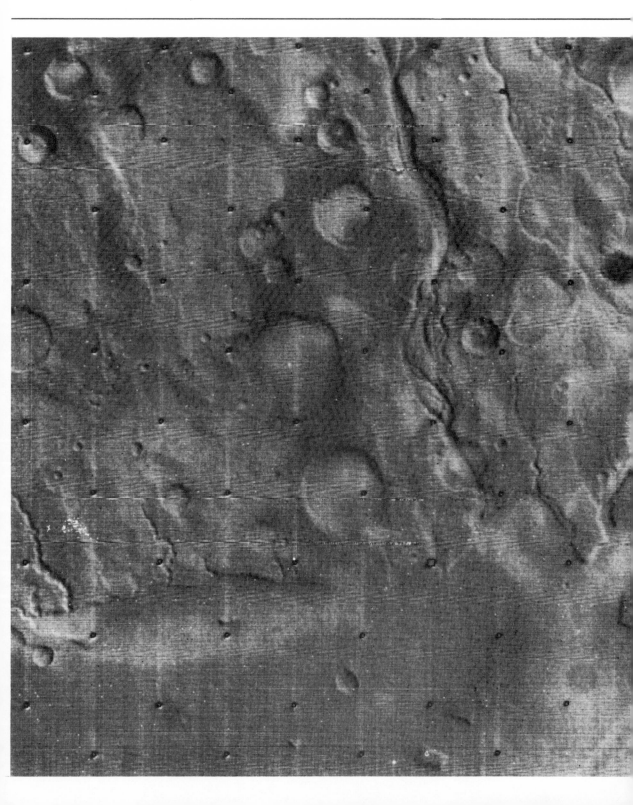

at the University of Michigan and a sometime geologist. McLaughlin argued that there were great volcanoes on Mars and that windblown volcanic ash was responsible for the Martian dark areas and their time variations. While some of the details of McLaughlin's proposal do not survive today's scrutiny, he was remarkably prescient in deducing two of the major features in the Martian environment from an extreme paucity of data. Or maybe he was just lucky. Occasionally someone has to guess right!

The Mars revealed by Mariner 9 is meteorologically, geologically, and just possibly biologically much more interesting than many scientists had previously suspected. But Mariner 9, despite its enormous quantity of data, has viewed Mars only from a narrow perspective, just as the Earth viewed by Apollo from Earth orbit gives little hint at the existence of hills and streams and trees—to say nothing of mice and microbes. In the same way, Mariner 9 gives little or no information about the close-up character of the Martian surface. Such an exploration requires a landing mission. The United States has plans well under way—if they are not canceled in some terminal mania for throwing out the scientific baby with (or without!) the military-industrial bathwater—to land two spacecraft on Mars in the summer of 1976. July 4, 1976, is a possible and therefore probably an inevitable date for the first landing.

These Viking spacecraft are remarkably sophisticated combinations of scientific instruments, designed to examine Mars for microorganisms, organic chemistry, surface mineralogy, winds, Marsquakes, magnetic dust, exotic atmospheric gases, and a wide range of other phenomena. They will return panoramic color photographs of the surface of Mars at two sites. The preliminary plans are for the first Viking lander to make Martian landfall in a region called Chryse—the land of gold. It is a place which appears to be low enough so that the Viking aerodynamic braking system will work, smooth

MORE MYSTERIOUS CHANNELS
This view shows more of the mysterious channels that have attracted so much attention concerning possible aqueous epochs in the history of Mars. These are incised in an older eroded, cratered terrain. The view is about 300 miles wide. The lighting is from the left.

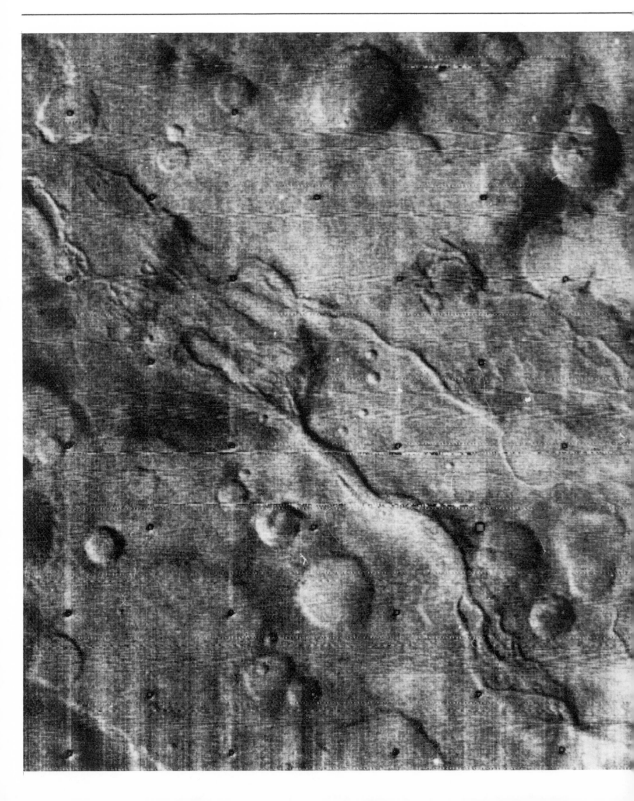

enough so that the lander will not tumble over, generally wind-free enough that it will not be blown over, and soft enough so that there will be material for the automated scoop to scoop up. It is also—by great good fortune—also scientifically interesting. For Chryse is a region of sinuous tributaried channels—the presumptive relics of a past epoch of running water and clement conditions on Mars.

The Soviet Union will probably attempt a duplication of its Mars 2 and 3 efforts in 1973—a time unlikely to have a global dust storm and therefore much more propitious for success. The first surface photographs of Mars may come from a Soviet lander. But actual life-detection experiments by the Soviet Union are unlikely before the 1976 opportunity. We are thus on the doorstep of another epic phase in our exploration of Mars. No one knows what the Viking and comparable Soviet landers will uncover. But if Mariner 9 is any guide, astonishments and delights and high scientific adventure are in the offing.

Mars and its moons are only a small sampling of the nine planets, thirty-two moons, and innumerable asteroids and comets which fill our solar system. There is our nearest planetary neighbor, Venus, a hellhole of a world which may, however, offer a cautionary tale for the evolution of the Earth. There is Mercury, a place of very high density, possibly a world with its crust and upper mantle stripped off in the early history of the solar system. Beyond Mars are the Jovian planets, which dominate our solar system as seen from afar. Almost all the mass and angular momentum of the solar system resides in Jupiter, Saturn, Uranus, and Neptune. These giant planets have retained the hydrogen-rich gases of the early solar system—the gases from which life on Earth evolved. And the comets may be the most pristine samples available to us of the material which formed the solar system. They are also fundamentally

STILL MORE CHANNELS
A continuation of the pattern of sinuous channels seen in the preceding photograph.

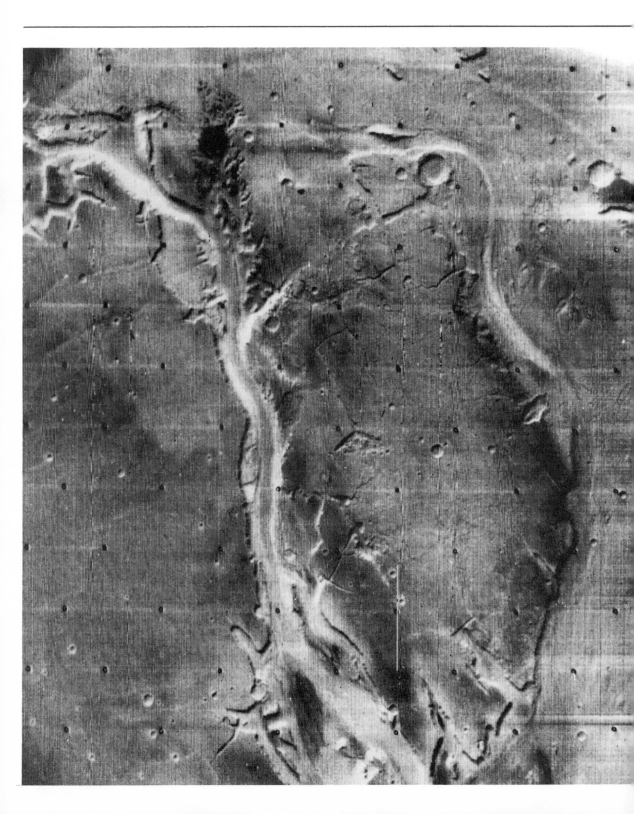

interstellar objects, spending the vast bulk of their lifetimes out in the dark between the stars.

Within this vast array of worlds in our solar system—which we already know to be a fascinating collection—there are bound to be surprises. One recently emerging surprise is Titan, the largest satellite of Saturn, a world about the size of the planet Mercury. Recent work shows that Titan has a dense atmosphere, deep-red clouds, and a surface temperature much higher than it should have for its great distance from the Sun. Titan is about ten times farther from the Sun than the Earth is and receives about 1/100 of the sunlight, yet its surface seems to be almost twice as hot as an object that far from the Sun should be. The explanation seems to be the greenhouse effect, which shuts in the thermal infrared radiation given off by the surface of Titan. The agent responsible for the Titanian greenhouse effect appears to be molecular hydrogen, the most abundant molecule in the universe. The surface atmospheric pressure may be a few tenths that on Earth, and although Titan has a much denser atmosphere than does Mars today, the gravity of Titan is weak enough to allow the hydrogen to escape rapidly into interplanetary space.

The hydrogen escaping from Titan is probably blown backward from the Sun by the solar wind, and in a strange sense Titan may be a comet of immense proportions. The density of the solid body of Titan is low, approximately two grams per cubic centimeter—about halfway between rock and water. The interior of Titan probably contains snows and ices of water, methane, and ammonia—the very same ices that are thought to compose the comets.

The Jovian planets themselves probably are similarly loaded with organic compounds, and the entire outer solar system may be a large-scale natural laboratory that has been working on the chemistry of the origin of life over the last 5 billion

AND MORE
In this view a smooth plains area has been incised by a sinuous channel. The entire view is about 400 miles wide.

years. But the gravities on Jupiter, Saturn, Uranus, and Neptune are so high as to make very close approaches or landings on them impractical, at least for the near future. Titan, however, is a much more accessible object and if we are wise enough to use our resources properly, will probably be the first object rich in such organic matter to be investigated. The two projected United States Mariner Jupiter/Saturn flybys—two vehicles to be launched in 1977—will arrive near Jupiter in 1979 and Saturn in 1981. The targeting of these space vehicles is now sufficiently in hand so that one of them could fly within a few hundred miles of the surface of Titan, there to examine the blood-red clouds, the atmospheric composition, and the as yet unseen surface by instruments of the sort proved out by Mariner 9.

But like Viking, Mariner Jupiter/Saturn is in financial trouble. Yet by many standards, such missions are inexpensive. Mariner Jupiter/Saturn costs about the same as the American aircraft shot down in Vietnam in the week in which I am writing these words (Christmas 1972). The Viking mission itself costs about a fortnight of the Vietnam war.

I find these comparisons particularly poignant: life versus death, hope versus fear. Space exploration and the highly mechanized destruction of people use similar technology and manufacturers, and similar human qualities of organization and daring. Can we not make the transition from automated aerospace killing to automated aerospace exploration of the solar system in which we live?

The advantages of such exploration are varied and, to me, compelling.* I believe that the scientific perspective obtained by observing our neighboring worlds in the solar system will have major practical benefits back here on Earth, in addition

---

* I have discussed them in more detail elsewhere. See Carl Sagan, *The Cosmic Connection* (New York: Doubleday, 1973).

EROSIONAL DETAIL

A long scarp is being eroded as well as some branching tributaries in this high-resolution view. The lighting is from the left and the width is about 300 miles.

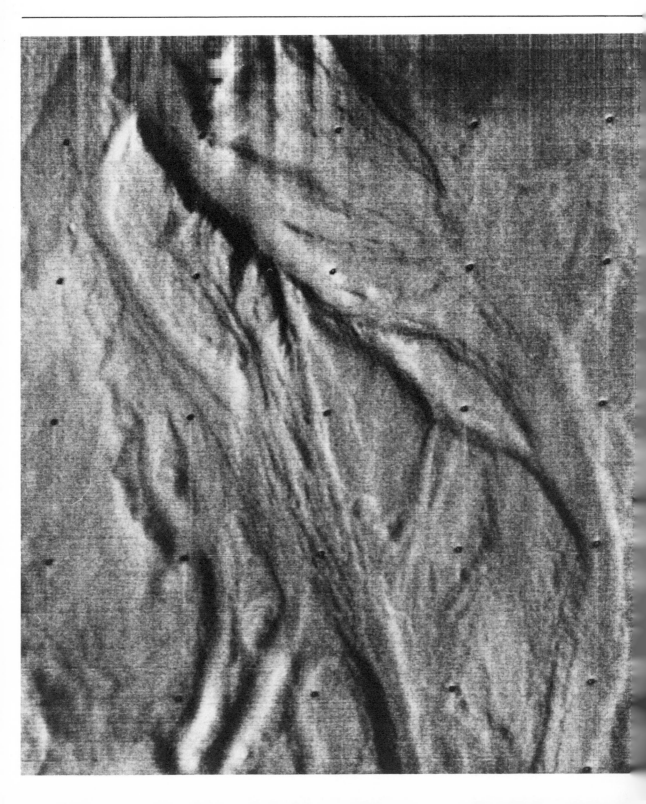

to providing a sense of peaceful adventure, of the exhilaration of exploration, at a time when all the Earth's surface has been explored. When our sciences of meteorology, geology, and biology are generalized by contact with other examples elsewhere, their powers will be vastly enhanced. Space exploration also provides a new perspective on our own planet, its origins, and its possible futures. We see the Earth as it is, one planet among many, a world whose significance is only what we make it. We realize that if there is life elsewhere, it will almost certainly be very different from life on Earth, and this makes the similarities among men clear and awesome compared to their differences.

There is a great need for social reform on Earth, for the removal of poverty, starvation, and injustice. But in addition to food for the body, we need food for the mind and spirit. As I read human history I find a remarkable correlation between epochs of exploration and discovery and epochs of major cultural advances. By the exploration of the solar system we find out, and make better, who we are.

BRAIDED CHANNELS
The pattern of overlapping irregular channels and bars shown here is quite similar to that produced on Earth by intermittent large-scale flooding, and is perhaps the most convincing evidence so far for the past presence of flowing water on Mars. Nevertheless, the origin of these features remains a matter of speculation. The view is about 30 miles wide.

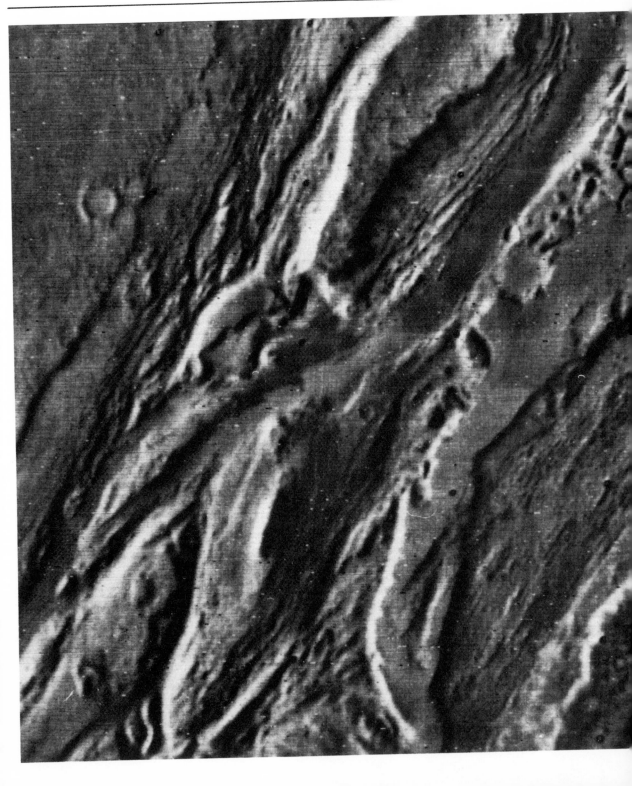

Walter Sullivan

Mars, until we came to know it close at hand, was for much of mankind a dreamworld. Although many people no longer quite believed it to be inhabited by Martians, they like to think it might be. To them the reality of Mars is a letdown. Yet anyone who peruses the pictures in this book cannot but become extremely curious as to how those extraordinary features came about. There are awesome canyons, including the greatest such feature known anywhere; snaking "rilles" like those in some parts of the moon; irregular regions of grossly subsided ground; and terrain cut into squares by a hatched pattern of ridges, like hedgerows on an English landscape. There is a crater whose inner walls seem almost vertical, surrounded by a smooth region showing no evidence of the eruption or impact that could have formed it.

These features set Mars distinctly apart from the only other celestial bodies whose surface features we know in detail: the Earth and Moon. On Earth there are mountain ranges that extend for thousands of miles, manifesting the activity that constantly spreads apart the ocean floors. If the oceans dried up, the long mid-ocean ridges from which this spreading occurs would appear, as would the trenches, flanking continents (like South America) or island arcs (like Japan), where the moving sea floor descends back into the Earth's interior.

On Mars we see only hints of such activity. Its great equatorial canyon is wide enough in places to mimic what on Earth would be regarded as an incipient ocean basin, with a ridge down part of its centerline. In such respects, Mars is not entirely an alien body. Its features are at least close enough in appearance to those on earth so that we can put a name to them. When we get our first close looks at the planets beyond Mars, from Jupiter on out, we can expect no such familiar features. We will need an entirely new vocabulary.

117

AND STILL MORE CHANNELS
This photograph and the two succeeding ones show high-resolution views of a series of sinuous and to some extent braided troughs, again suggestive to some of the action of liquid water. The lighting is from the left, and the area in each picture is about 30 miles wide.

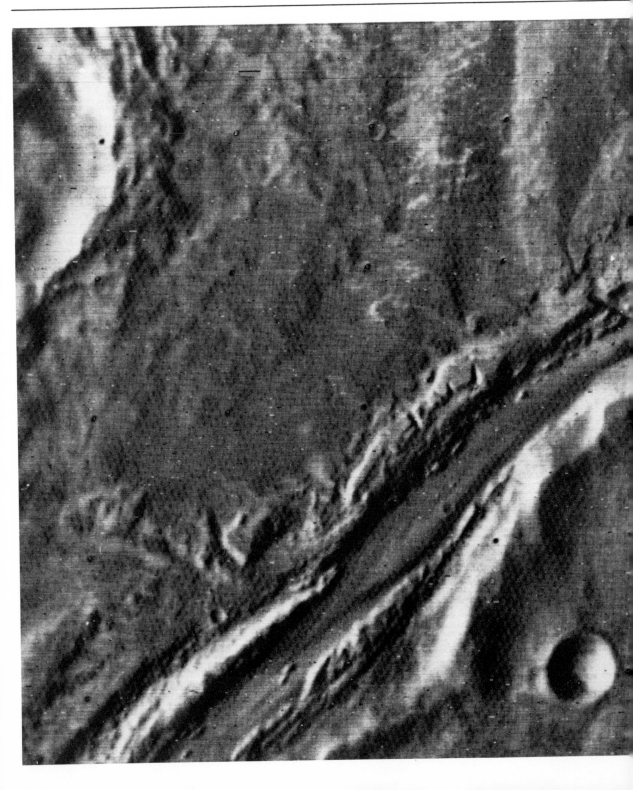

Having had our first close-up views of the Martian reality through the eyes of Mariner 9, we cannot stop there. Two opportunities stand immediately before us. One is to use the tried-and-true Mariner system for similar surveys of other planets. Venus is closer, though its seemingly unbroken cloud cover will make photographic missions less informative. Some believe that Mercury resembles the Moon in composition (Mercury is less than twice the diameter of the Moon) and that close-up views would have much to show about the history of the planet that has been closest to the sun over the past 4.6 billion years.

While Jupiter is largely cloud-covered, the clouds are organized into bands and amongst them is the great Red Spot, one of the most puzzling features of the solar system. What a thrill it will be for those who watch as a Mariner-type system prints out, bit by bit, the first detailed views of this region!

More Earthlike features can be expected on some of the larger moons of the outer planets—one, at least, seems to have a surprisingly dense atmosphere.

The other immediate challenge is, of course, to land on Mars—first with unmanned vehicles partially controlled by radio command from Earth but also with a certain computer-stored "intelligence" to permit on-the-spot reactions to situations requiring quick response. Otherwise, an exchange of signals with a spacecraft on the Martian surface, even when Mars was relatively close, would take at least ten minutes and, when Mars was on the far side of the sun, the round-trip travel time for such signals would be forty minutes or more. And this could mean that the response from the mission controllers on Earth would be too late to save the mission from some mishap or to take advantage of some fleeting observation opportunity.

Such an "intelligent" automaton may eventually be succeeded by men, although the voyage to Mars, lasting many

A continuation of the preceding view.

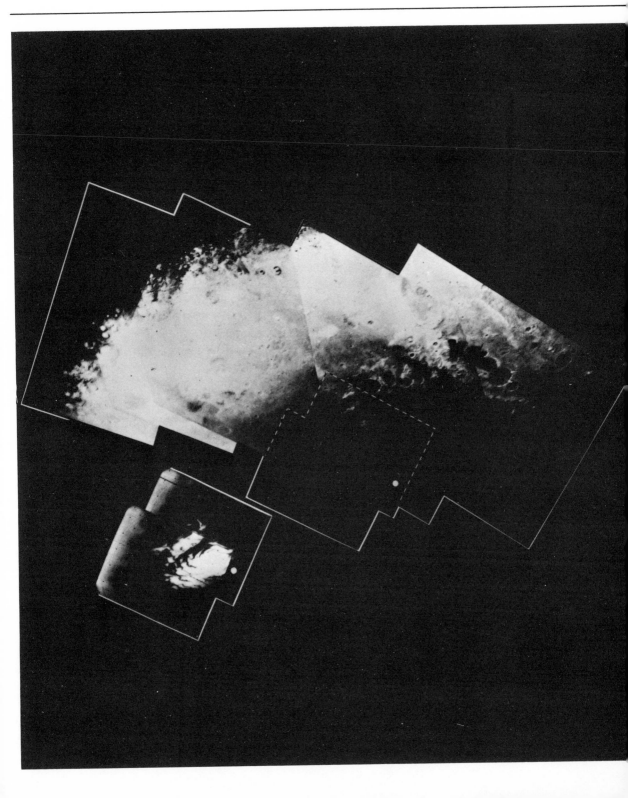

tedious months, could be almost unendurable unless, perhaps, drugs were used to induce prolonged sleep.

Ultimately, our "educated" automatons should qualify for journeys far out into the solar system and beyond. There, the distance from Earth will make it even more essential that they "think for themselves." One can only speculate as to what these messengers will tell us, for the realms that they could reach lie far beyond our direct ken. Jan H. Oort, the Dutch astronomer, has proposed that the solar system is encircled by a cloud of 100 billion comets, moving around the sun in very slow orbits as a frothy, frozen residue from the formation of the sun, planets, asteroids, and meteorites. This cometary zone would be from 100,000 to 150,000 times farther from the Sun than is the Earth. This is roughly two light-years, which would place it close to the boundary region where the domain dominated by solar gravity gives way to the gravitational realm of other nearby stars.

Will it be possible to obtain pictures and scientific data from a spacecraft that far away? Signals were obtained from Mariner 9 when it was orbiting Mars on the far side of the Sun, almost two and one-half times the Earth-Sun distance, and if one is willing to accept slow transmission rates—days, perhaps, to send a single picture—far greater distances should be feasible.

Further down the line is the possibility of sending such craft to visit the nearest planetary systems. Despite the deep human desire to know if there are other worlds like our own, we have firm evidence of only one planetary system apart from our own—that which orbits Barnard's Star, six light-years away. We cannot see the planets themselves, but we can detect their gravitational effect on the motion of that star: the star's track, against the backdrop of very distant stars and galaxies, deviates from a straight line as the gravity of these orbiting planets pulls the star slightly to one side or the other.

It is because Barnard's Star is so close that we can see this

THE SOUTH POLE OF MARS

The upper mosaic of Mariner 7 pictures shows the south polar region as it appeared in 1969, and the small inset shows the small residual cap as it was observed by Mariner 9 when it first arrived at Mars in November 1971.

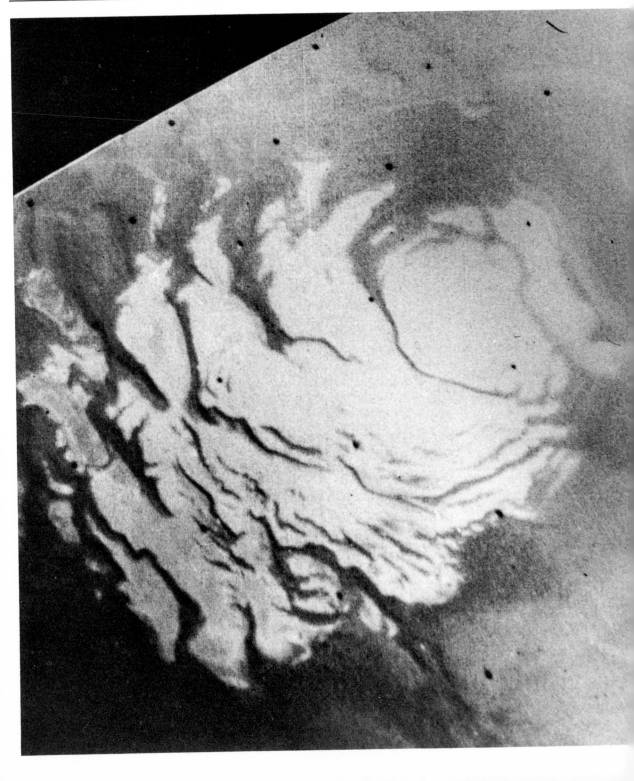

effect, and there is good reason to believe that planetary systems are common among the billions of stars forming the Milky Way galaxy. The three-star system of Alpha Centauri is closer than Barnard's Star, but stable planetary orbits are unlikely where the gravity of three stars is competing for control over a planet.

For our automated explorer to reach the vicinity of Barnard's Star would take a sizable portion of one human lifetime, even with rocket systems more sophisticated than any available today. The announcement of its discoveries, traveling back to us at the speed of light, would take six years to get here. But if we can find ways to keep in touch with a messenger out in Oort's comet zone, two light-years away, we eventually should be able to reach across distances several times greater.

To what extent will the public be willing to pay for such ventures? Man has an inborn curiosity about what lies over the hill or around the corner. It becomes manifest as soon as a baby can crawl. It is characteristic, to some extent, of all mammals, and when combined with man's intelligence it becomes what we call intellectual curiosity.

But if one does not see the hill or the corner that tempts our curiosity, the urge to go and see is far less or even nonexistent. The Moon is the most highly visible object in the skies, apart from the sun, and its surface features, broadly speaking, can be seen with the naked eye. It was therefore relatively easy to arouse public interest in Project Apollo. With a television camera mounted on the Lunar Rover, under the control of someone in Houston who turned it this way and that to follow the activities of the astronauts on the Moon's surface, it was easy for those on Earth to identify with those two men, in their inflated space suits, kangarooing across the Moonscape. But one part of the Moon looks a good deal like another, and once the novelty wore off, public interest in the project sagged. As attention turned increasingly to social problems at home, there

THE SOUTH RESIDUAL CAP
This computer-rectified picture of the south polar cap was taken a month after Mariner 9 went into orbit and showed the characteristic pattern of bright circular frost. The cap was about 200 miles wide.

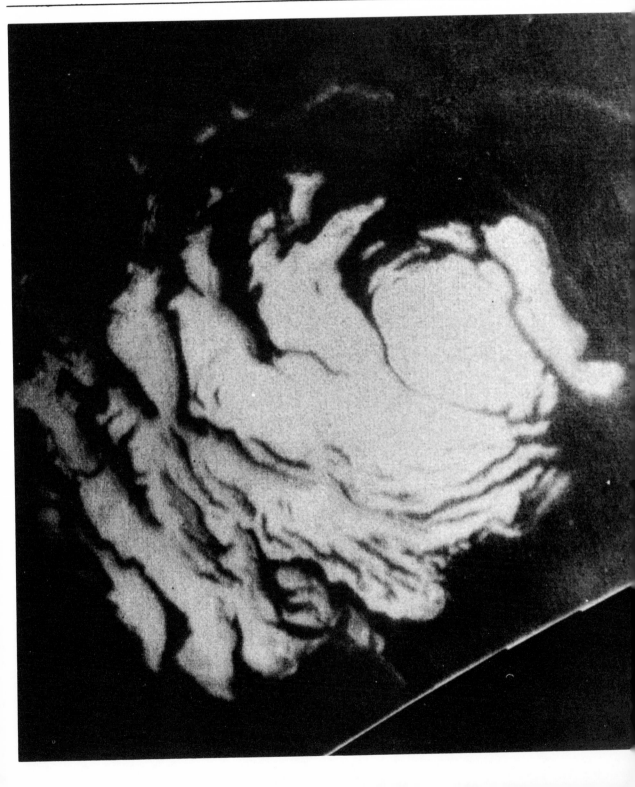

seemed to be more grumbling about the cost of the project than enthusiasm for its discoveries and the intellectual challenges they presented to those seeking to explain them.

It may therefore be some time before the public is willing to support manned expeditions beyond Earth orbit. But unmanned missions—even those of a relatively ambitious nature—can be undertaken for far less annual cost. Apollo demanded such a far-flung and elaborate system of ground stations, biomedical monitoring facilities, and the like, that to keep the organization intact was very expensive and made it uneconomical to slow down the program below one or two shots a year.

Unmanned missions are far less demanding in these respects and can be conducted at a more leisurely pace. But unless the public sees "the hill," there will be no interest in climbing it. Hence, the future of space exploration will depend in great measure on the manner in which future generations are educated and kept informed.

We must also expect the pendulum swings of public interest to continue. In 1960, in the United States, there was concern that the country was becoming "second-rate" because of its failures in space technology. A decade later the focus was on the environment, pollution, and other matters. When the first Viking lander sets forth for Mars a few years hence, we can expect at least a partial revival of excitement, even if no one can honestly hold forth the likelihood of finding life there.

Of very great importance in the planetary missions will be the provision of good-quality photographs. The Earth-controlled "eye" on the Lunar Rover provided a valuable scientific record of what the astronauts were doing, both for instantaneous coaching by specialists and as a record for study later on. But its chief immediate value was in mustering the popular enthusiasm without which such costly efforts cannot be launched.

Every time we have lifted new instruments above the atmos-

THE SOUTH POLAR CAP AGAIN
This view of the cap was taken during the south polar summer, more than two months later in the mission than the preceding view. Surprisingly, the cap had not changed its outline significantly, which suggests the possibility that it is composed of water ice—not solid carbon dioxide, as are the large seasonal caps.

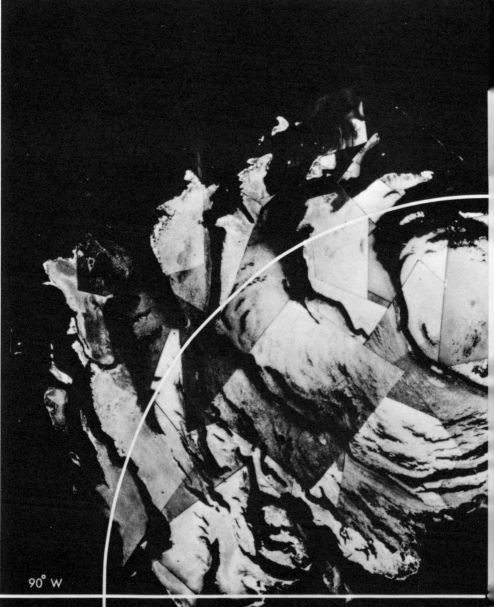

90° W

SOUTH P

phere, or looked about the universe at new wavelengths or with new sensitivities, we have made startling discoveries: objects emitting highly rhythmic radio pulses (at first suspected of being the creations of distant civilizations, but now believed to be rapidly spinning objects of great density) known as pulsars; objects seemingly farther away than anything else we can see and shining with almost unbelievable brilliance, known as quasars; objects that, at least superficially, seem to be flying apart faster than the speed of light (supposedly a physical impossibility); or something as close to home as the belt of trapped radiation that encircles the Earth.

As long as we remain below the obscuring blanket of air that covers the Earth (and shuts off much of the information-bearing radiation from beyond), or even as long as we refrain from venturing much beyond the vicinity of the Earth, our knowledge of the solar system, the galaxy, and the universe will remain provincial and limited. Yet the triumph of science and reason over superstition, whose first great landmark, five centuries ago, was the discovery by Copernicus of the planetary nature of the Earth, will not be complete until we have pushed our knowledge of the reality of the universe to the limits of our capabilities, both technological and intellectual.

We do know enough already, however, to believe that no myth or legend could be as rich in beauty, wonder, and awe as the full reality of the universe that is our home.

A CLOSER VIEW
A mosaic of high-resolution views of the residual south polar cap, showing characteristic dark bands delineating bright patches of frost. The dark bands are believed to be outward-facing steps separating patches of white frost located more or less on level ground.

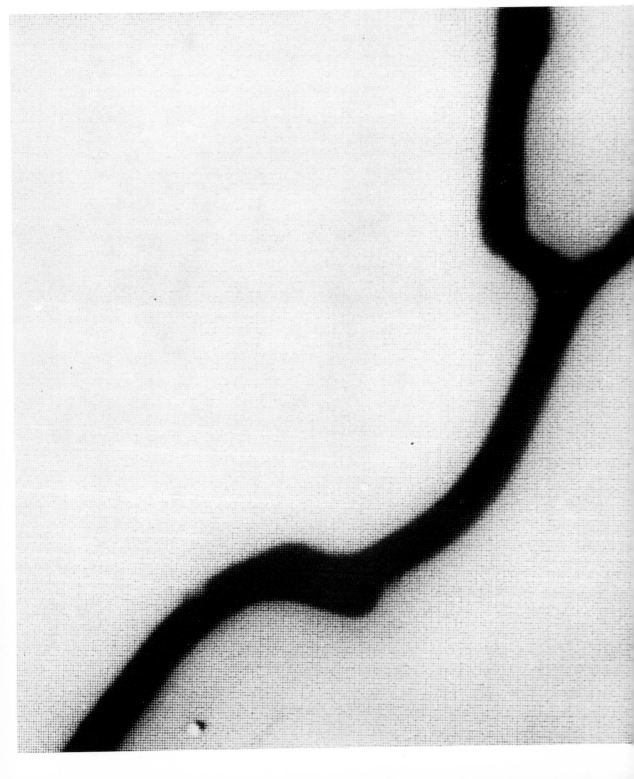

Ray Bradbury

Rereading our words from more than a year ago, I am filled with alternate lifts of elation and mirror-image semidepressions. Mostly, however, I wish to explode, which has been my tendency since Creation first tucked its dynamite cap in my backside and kicked me into the world.

I have all sorts of *non sequitur* reactions to the fascinating but calmer words, and the later essays, of Messrs Clark, Sullivan, Murray, and Sagan. I find myself speaking in tongues lately. That could easily mean that the approach of poetry simply signals my incipient senility. Be that as it may, here is my first response to the text you have just finished:

What I to apeman
And what then he to me?
I an apeman one day soon will seem to be
To those who, after us, look back from Mars
And they, in turn, mere beasts will seem
To those who reach the stars;
So apemen all, in cave, in frail tract-house,
On Moon, Red Planet, or some other place;
Yet similar dream, same heart, same soul,
Same blood, same face,
Rare beastmen all who move to save and place their pyres
From cavern mouth to world to interstellar fires.
We are the all, the universe, the one,
As such our fragile destiny is only now begun.
Our dreams then, are they grand or mad, depraved?
Do we say yes to Kazantzakis, whose wild soul said,
God cries out to be saved?
Well, we go to save Him, that seems sure,
With flesh and bone not strong, and heart not pure,
All maze and paradox our blood,
More lost than found,
We go to marry stranger-flesh on some far burial ground
Where yet we will survive and, laughing, look on back

129

SOUTH POLAR CAP DETAIL
This view of one of the dark bands in the preceding frame corresponds to a locality in the upper right-hand corner of that frame. It is about 50 miles wide and shows a surface entirely coated with the white carbon-dioxide annual frost, execpt for the dark, outward-facing slope.

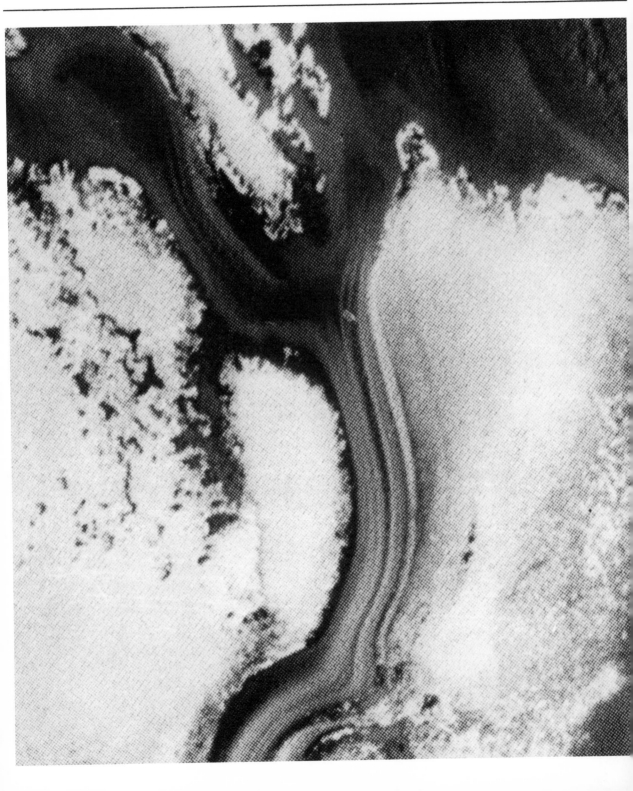

To where we started on a blind and frightful track
But made it through, and for no reason
Save it must be made, to rest us under trees
On planets in such galaxies as toss and lean
A most peculiar shade,
And sleep awhile, for some few million years,
To rise again, fresh washed in vernal rain
That is our Eden's spring once promised,
Now repromised, to bring Lazarus forth forever,
Stoke new lamps with ancient funeral loam
To light cold abyss hearths for astronauts to hie them home
On highways vast and long and broad,
Thus saving what? who'll say salvation's sum?
Why, thee and me, and they and them, and us and we . . .
And God.

Immediately, one almost has to leap into a wild Scots sword dance around, about, over, and down on that loaded word: God.

For if you think I am talking about that dreary old bore of a God who sat around writing up life transcripts and noticing dead sparrows, by the bowels of Darwin, no!

I trot with Kazantzakis, who has written the best new New Testament of our time, *The Saviours of God,* a book that will probably be transported as A-Plus V.I.P. literature in every rocket for the next seventy thousand years or unguessed calendars beyond.

I run with Bernard Shaw, who long before most of us were born brooded on matter and force making itself over into intelligence, imagination, and will.

And the text of this book that you hold in your hands only makes me trot and run harder to examine the possibilities of good men trying to be better.

The universe is full of matter and force. Yet in all that force, amongst all the bulks and gravities, the rains of cosmic light,

SOUTH POLAR CAP DETAIL REVISITED

This photograph, of the same area as that shown in the preceding view, but taken some months later, shows the appearance of the south cap at high resolution when all the carbon-dioxide has sublimed, leaving a residual cap—probably of water ice—along with bare ground. The peculiar "dark bands" in previously dark areas correspond to layers of differing slopes and reflectivity, a characteristic polar formation on Mars. These layered deposits entirely surround and include the residual caps at both poles.

the bombardments of energy—how little spirit, how small the decimal points of intelligence.

Dumb, sometimes—yes. Awful, quite often. Dreadful apish brutes on occasion following occasion. That's how we things that represent intelligence seem to ourselves, and quite often truly are.

And yet I would not see our candle blown out in the wind. It is a small thing, this dear gift of life handed us mysteriously out of immensity. I would not have that gift expire. Crossing the wilderness, centuries ago, men carried in covered cows' horns the coals of the previous nights' fires to start new fires on the nights ahead. Thus we carry ourselves in the universal wilderness and blow upon the coals and kindle new lives and move on yet once more.

If I seem to be beating a dead horse again and again, I must protest: No! I am beating, again and again, living man to keep him awake and move his limbs and jump his mind.

For the absurd thing is we *do* get so busy with facts, with looking at photographic plates, with calm theories, and with everyday life that we forget the reasons behind it all.

I was reminded of this the other morning when I got out of bed and suddenly thought: Why, sweet Jesus, what's the use of looking at Mars through a telescope, sitting on panels, writing books, if it isn't to guarantee, not just the survival of mankind, but mankind surviving forever! Good God in heaven, we were born to live, and live in mystery, which crowds all about and would smother us if we let it.

Our situation reminds me of that remarkable Pirandello story in which an old man on a train describes the bravery of his slain son in war and talks about his medal, which he then shows and discusses by the hour. Finally someone on the train leans forward and says, gently, "Your son, then, is he *really* dead?"

POLAR LAYERED DEPOSITS
These striking layers are part of an extensive area surrounding both poles of Mars. They have very smooth surfaces and very few impact craters, and the characteristic steps are associated with the slopes of each layer. The light is from the left, and the area shown is about 40 miles wide in this frame.

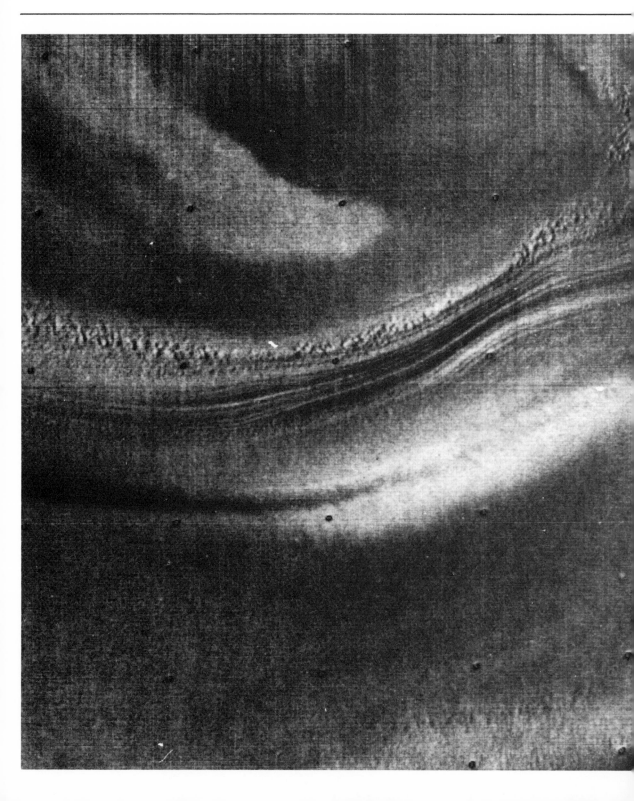

The old man collapses into sobbing. He is destroyed. Until this moment he has not truly known the death of his son.

This book is a corollary to that tale. Looking at the photographs, reading the words, I say to myself, I turn and say to *you:*

You, then, are you *really* alive?

This book, and everything in it, is useless unless considered as one more stone put forth to step on as we move across a river so dark, so deep, that one error would drown us forever.

The photographs you find here are pictures of your next home. Know and learn them well. You will be living there a long time. By you, of course, I mean mankind/aggregate-in-motion.

These Mariner photographs *do* look lonely, don't they? Far lonelier than Barsoom ever seemed, and most certainly hostile.

One of the ideas that Dr. Krafft Ehricke tossed forth during an afternoon's conversation in San Diego some thirteen years ago concerned itself with man striking solid rock on Mars, even as Christ struck boulders and stones, to bring forth liquid miracles. That is to say, build such artificial volcanoes, superb factories of fire, as can boil rocks to lava and set free those chemicals locked in stone that can be breathed as oxygen, or drunk as combinations-made-water.

A titanic experiment, to say the least. But man, with blueprint and Vulcan's blowtorch, can mouth-to-mouth breathe Mars into new existence.

Some of you will immediately say we go to pollute Mars.

You are the people who see a partially filled glass as half empty.

I see the glass as half full.

I say we go to save Mars from itself.

And do ourselves favors, meanwhile.

MORE LAYERED DEPOSITS
Another view of the characteristic polar terrains.

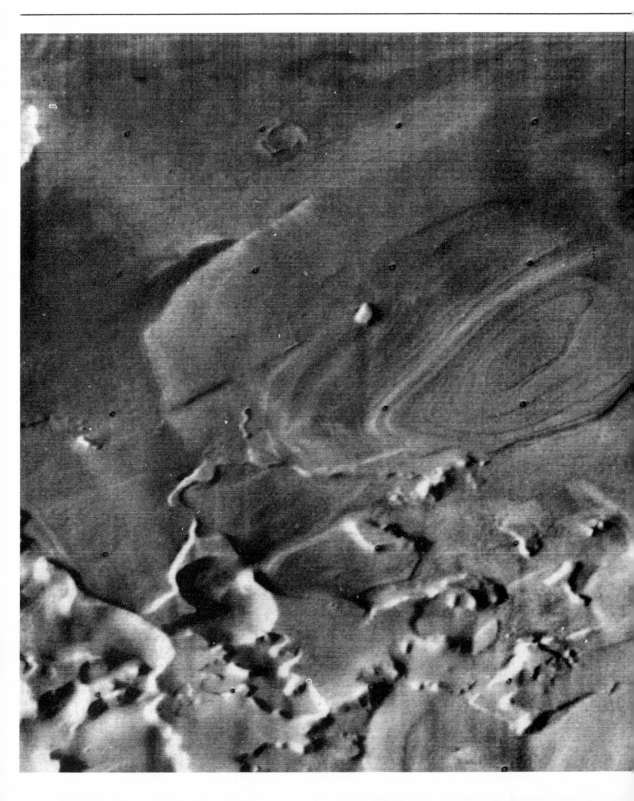

Paradoxically stated: what is not polluted is elevated. I live inside that last word.

Are we, therefore, grand, great, good tall heroes deserving of fame and immortality? Hardly, as my poem at the start of this epilogue surely indicates. But, also, we are *not* villains, as many prolapsed intellectuals of both left and right have wailed the past few years among other years. We are poor beggars in the long night of the abyss, begging for crumbs on cold street corners where death is certain if once we stumble. Are we beautiful, lovely, endearing, wondrous models of morality? No. We are Quasimodo, ten billion times squared, hunched of back, blind of eye, pitiful of stance, yet reaching to pull that rope and ring all the loud bells of the universe and listen to them—forever. And we shall do it.

For the dream of mankind has been to someday kill death. We have written of it in our stories, novels, songs, poems. Dylan Thomas says: "And death shall have no dominion." John Donne concurs: "Death shall be no more; death, thou shalt die."

We echo them and cry out to the Reaper: Beware of our rocket, which will shatter your scythe and scatter its bits to the stars.

Many thousands of years from this day, how will we name this time we lived in? Exodus II? Genesis Revisited?

Ecclesiastes Rediscovered might be better. If for all things there are seasons, times to come and go, live and die, then surely ours is the Time of Going Away, a very special long season of knowing and doing.

Man should not be diminished by his hostile activities but replenished by the thought and imagination of such men as sat onstage with me at Caltech in November 1971 and drummed up their hearts and gathered their wits to try to figure just what in Creation's name it all meant.

THE EDGE OF THE POLAR LAYERED DEPOSITS
This frame shows the southern boundary of eroded polar layered terrains adjacent to smooth plains seen in some previous views.

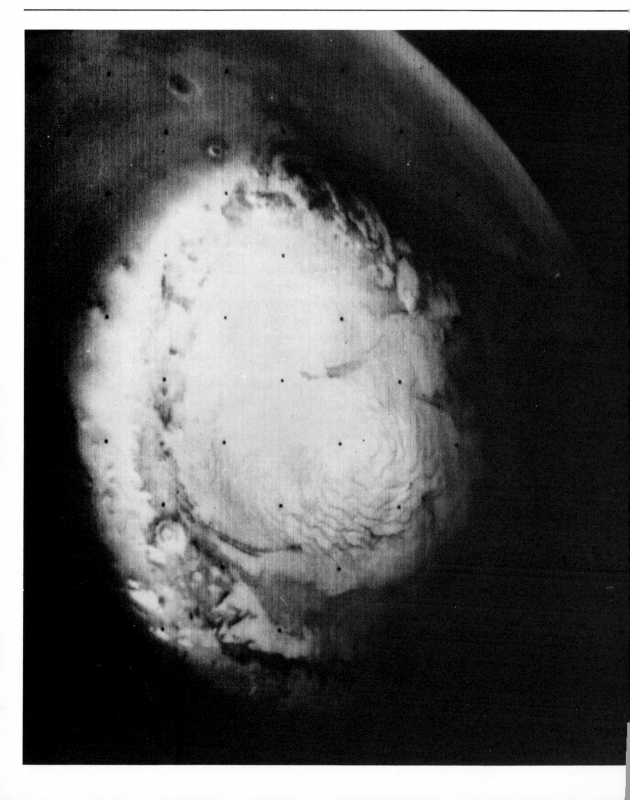

They came, as I came, to familiarize themselves with new events, new facts.

Familiarize. *There's* a word to play with. There's something to word-associate around.

The word *familiar* conjures up that process whereby we combine strange facts together into families so that we forget how strange they still remain.

It is the duty of the sciences to break down the barriers between families of knowledge every few years so that we resight, realign, reexperience the miraculous-strange and re-combine its components into new families.

And we do this not only with new data collecting but new emotionalism.

The most common sound from the Moon, as we explore it, is the cry of the technician/scientist shouting in surprise: Look here! There! See this! See that! Good God! Oh, Jesus!

These are not blasphemous but celebratory cries.

These are the same sounds heard from cathedrals and synagogues and Moslem squares at sunset. It is the gasp of the artist discovering beauty. It is the insuck/output of the aesthete as well as the physicist. We all have strange awe in us; we put it forth with dissimilar motives but similar breaths.

We live in miracles which cannot be explained. The scientist, the theologian, the artist—each attempts impossible explanations. That is why we love and admire and hope for them so. We wind up with theories from each for making do. We have been busy with that game, intuited from our marrow. Whatever blew its fiery breath into us, as into gloves, 3 billion years ago, exhaled in that moment a gigantic whisper which has touched our multitudinous ears ever since, and that whisper says: Sweet man, dear blood, wild creature, rare device of the universe, fragile flesh—survive.

THE NORTH POLAR CAP
This frame, taken late in the mission, shows the retreating north polar cap.

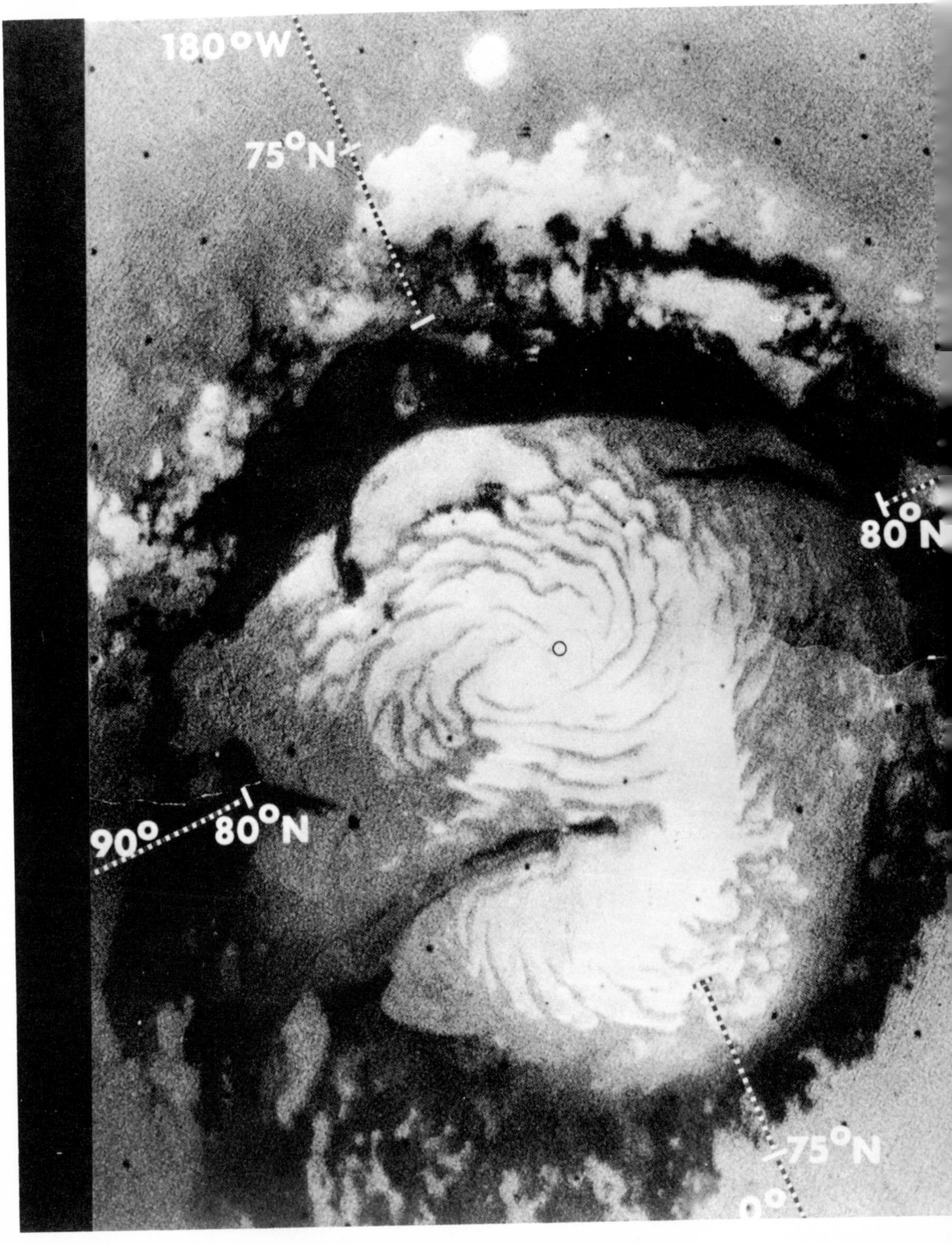
180°W
75°N
80°N
90°
80°N
75°N

We heard that whisper from the Moon. We hear Mars calling with an even louder voice, if we tune up our hearing.

At such times it is dreadful to recall all the times in the past five years when I have heard NASA apologists applying poultices to the tax wounds of the American public by describing how the Apollo missions have given old ladies new plastic cooking pans made from nose-cone chemistries, and attached more proficient bumpers to mindless autocars. With thinking like that, one cannot set out to rape an amoeba, much less touch the Moon.

Toynbee speaks of the challenge and response of various tribes, nations, and racial groups in the long history of man. Those who refuse the challenge, who will not respond, become the detritus of history. The universe will not accept mediocre lunacy save to tread upon it, grind it under, and go on to other yeasting experiments.

I wonder how many of you recall that vivid and exhilarating motion picture by H. G. Wells *Things to Come,* wherein a mixed mob of intellectuals rushes to prevent the takeoff of a moon rocket. Their leader, played by Sir Cedric Hardwicke, shouts:

"We don't want mankind to go out to the Moon and the planets. We shall hate you more if you succeed than if you fail. Is there never to be calm and happiness for man?"

To which Raymond Massey, playing Cabal, the instigator of space travel, radios his reply:

"Either life goes forward or it goes back. Beware the concussion!"

The rocket fires.

In a vast telescope mirror, the fathers of the two astronauts watch the small light of the rocket moving toward the Moon, and one speaks: "My God, is there never to be an age of happiness? Is there never to be rest?"

To which Cabal answers: "Rest enough for the individual man. Too much of it and too soon, and we call it death. But

THE RESIDUAL NORTH POLAR CAP
This view, which has been stereographically rectified by a computer, shows Mariner 9's last view of the north polar cap, when it was nearly at its minimum size. The characteristic dark bands that were seen delineating the frost in the south are seen here as well, along with a large area of extensive white deposits apparently thick enough to cover over the dark bands. That area is believed by some to be the site of a large amount of excess solid carbon dioxide.

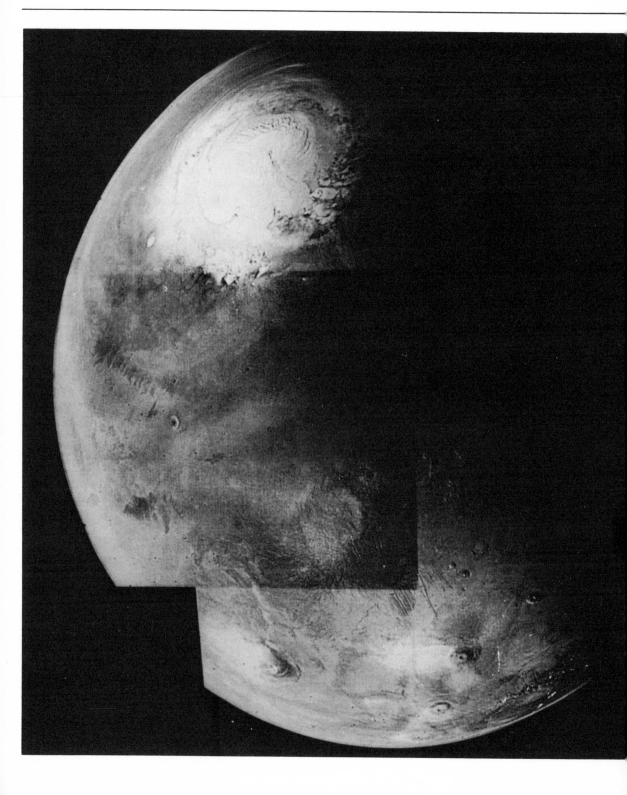

for *man,* no rest and no ending. He must go on—conquest beyond conquest. This little planet, its winds and ways, and all the laws of mind and matter that restrain him. Then the planets about him, and at last out across immensity to the stars. And when he has conquered all the deeps of space and all the mysteries of time, still he will be beginning."

He points out at the universe.

"It is that—or *this.* All the universe—or nothingness. Which shall it be?"

The two men fade. The stars remain. The music rises.

"Which shall it be?" his voice repeats.

So this book ends.

And man begins.

And this book has been a speaking of tongues and the tongues say Mars. I would love to be around when we make landfall there, so I beg the tongues to hurry in their romancing, and the corporate scientific eyes with their data collecting—for the sooner the dream, the sooner the foundation beneath the dream and the sooner men will seed themselves to Martians before our astonished gaze.

Get along to Mars and beyond.

The journey is long, the end uncertain, and there is more dark along the way than light, but you can whistle. Come with me by the wall of the great tombyards of all time which lie a billion years ahead. What shall we whistle as we stroll in our rocket, hoping to make it by the vast darkness where shadows wait to seize and keep us?

Follow me.

I know a tune.

Here . . . *listen.*

THE LAST LOOK

This mosaic of wide-angle frames returned by Mariner 9 late in the mission shows the north polar cap and a view of the planet all the way down to the large volcanoes and even the edge of the great canyon, as seen in the lower left-hand corner. Compare this photograph with the first one, opposite page 47.